ROYAL HISTORICAL SOCIETY

STUDIES IN HISTORY

New Series

TIME AND THE FRENCH REVOLUTION

THE REPUBLICAN CALENDAR, 1789–YEAR XIV

This series is supported by annual subventions from the Economic History Society and from the Past and Present Society

ECONOMIC HISTORY SOCIETY 1926 75 YEARS

PAST & PRESENT
a journal of historical studies

TIME AND THE FRENCH REVOLUTION

THE REPUBLICAN CALENDAR, 1789–YEAR XIV

Matthew Shaw

THE ROYAL HISTORICAL SOCIETY
THE BOYDELL PRESS

First published 2011

ISBN 978–0–86193–311–2

ISSN 0269–2244

A Royal Historical Society publication
Published by The Boydell Press
an imprint of Boydell & Brewer Ltd
PO Box 9, Woodbridge, Suffolk IP12 3DF, UK
and of Boydell & Brewer Inc.
668 Mt Hope Avenue, Rochester, NY 14620, USA
website: www.boydellandbrewer.com

A CIP catalogue record for this book is available
from the British Library

Printed in Great Britain by
CPI Antony Rowe Ltd, Chippenham and Eastbourne

FOR MY PARENTS

Contents

List of Illustrations

Table

Preface

This book is a history of the French republic's efforts to create a new division of time in the form of the republican calendar. It draws on extensive archival research to examine reactions to the republican calendar, its relationship to republican cultural reforms and to the longer-term context of early modern time-consciousness. It argues that the experience of time in the late eighteenth century was complex, and that the French were adept at working with several systems of timekeeping, whether that of the Church, civil society, or the rhythms of the seasons. This relationship was not unchanging. Developments in timekeeping technology, the spread of almanacs, clocks and watches, and changes in working patterns challenged early modern temporalities, and the new calendar may be viewed as a step on the path towards a more modern conception of time.[1] In this context, the creation of the calendar should be seen not just as an aspect of the republican programme of social, political and cultural reform, but as a reflection of a broader interest in time and the culmination of several generations' concern with how society should be policed.

Time and the French Revolution not only examines the effects the new calendar had on French society, but also seeks to widen historical understanding of the calendar from that which sees it simply through the lens of dechristianisation. The layering of eighteenth-century temporalities means that the calendar should not be examined using just the language of counter-revolution, Jacobinism or dechristianisation, but as an aspect of the state's wider involvement in everyday life and of the interplay between local politics and centralising processes. In many places, the calendar was followed alongside existing time codes, serving as one of the many new symbols of the republic; only occasionally did local practice provoke conflict and become, in the Jacobin gaze, a form of counter-revolutionary activity. The calendar is not just a historical curiosity, but had considerable contemporary consequences. In particular, this study compares the attempts to reform fairs and markets, official encouragement of the use of the new system of timekeeping, and the observance of republican festivals with earlier attempts to police timekeeping during the *ancien régime*, showing how what began as ideological concerns soon became administrative matters – a question of state-building and bureaucracy as much as revolution.

1 For a comparative study of shifts in time-consciousness see Hans-Joachim Voth, *Time and work in England, 1750–1830*, Oxford 2000.

Publication of this volume was aided by a generous grant from the
Scouloudi Foundation, in association with the
Institute of Historical Research

Acknowledgements

The research that underpins this book would not have been possible without the financial support of the British Academy and what is now the Arts and Humanities Research Council. I have also benefitted from research breaks granted by the British Library, and from the support from my section heads, C. J. Wright, Frances Harris and Carole Holden. The Library has also been a collegial place to think about times of change, and I am grateful for conversations with Andrea Clarke, William Frame and Jamie Andrews; in a reversal of the usual traditions of an acknowledgements page, I would also like to thank the readers of the British Library for always being a stimulating spur to discoveries in the collections. A debt of gratitude is owed to the staffs of the Bibliothèque historique de la ville de Paris, the old and new Bibliothèque nationale de France (and Gallica), the New York Public Library, the library of the College of the Holy Cross, Worcester, MA, the J. B. Morrell Library of the University of York, the Bibliothèque municipale and the Médiathèque Nelson Mandela, Besançon, the Hathi Trust and the archivists of the Archives nationales and numerous French departmental and municipal archives.

I would also like to thank the editorial board of the Royal Historical Society's Studies in History series and the executive secretary of the RHS, Sue Carr. I owe a debt to the anonymous readers at various stages in the writing of this book, as well as the support and suggestions of Rebecca Spang at a vital stage in its genesis. Rachel Hammersley, who witnessed much of the early research, ably and enthusiastically continued Rebecca's work as editor; I am similarly grateful to Christine Linehan for her work during the final stages of publication. They will not have saved me from myself, and the errors in these pages are, of course, entirely my own.

As many will attest, Alan Forrest is hugely generous with his time, store of knowledge and understanding of French history; I greatly appreciate this, as well as his rather wry delight in explaining some of the curious ways into the necessary archives and libraries. Colin Jones offered further encouragement and suggestions, and I hope that I have benefited from his ability to challenge assumptions. I would like to thank Martin Knudsen, Jill Maciak, Robert Poole, Mark Jenner, Chris Clarke and Geoff Cubitt for reading and commenting on various drafts, and Sanja Perovic, Richard Taws and William Nelson, who generously shared their own work on the calendar. David Garrioch offered early guidance via email as to where to begin to look in the

Paris haystack, leading me to a series of discoveries about markets, bells and policing. The European History 1500–1800 seminar at the Institute for Historical Research has been a vital forum for testing ideas over the years, as are the students of Birkbeck College, University of London. Here, I have much to thank Roger Mettam and Julian Swann for. Many of the chapters in this work have benefited from comments at seminars and conferences; Cynthia Bouton also acted as a model for a respondent at an early outing for my thoughts on rural opposition to the calendar, while David Andress and Ralph Kingston have fulfilled similar functions in a less formal context. Mark Bryant, John Langdon, Laurence Brown and Fay Bound Alberti remain model friends and historians; I would like to thank them and Elizabeth Cafferty for their support, good humour and encouragement over the years. They, like Daniel, Hannah, Marin, Fiona, Jamie, Matt, Brooke, Adi, Mike, Nina, Marie, the TroisV and, especially, Emily, each deserve their own *fête*.

Matthew Shaw
July 2010/Thermidor CCVIII

Abbreviations

AHR	*American Historical Review*
AHRF	*Annales historiques de la Révolution Francaise*
FH	*French History*
FHS	*French Historical Studies*
AD	Archives départmentales
AM	Archives munipales
AN	Archives nationales, Paris
APP	Archive de la Préfecture de Police de Paris
BN	Bibliothèque nationale, Paris

Introduction:
Writing the History of the Republican Calendar

The upheavals of the French Revolution gave the world many things. Perhaps the most mellifluous of these are the series of Latin and Greek-based names – *germinal*, *thermidor*, and so forth – chosen by the dramatist and politician Philippe-Franeois-Nazaire Fabre d'Églantine as the new months of the republican calendar.[1] These neologisms, with their linguistic allusions to the seasons and the agricultural year, remain inseparably linked to the period, dating such events as the fall of Robespierre on 9 *thermidor* or Napoleon Bonaparte's coup of 18 *brumaire*. In this fashion, the Revolution left its impression on time. Such an imprint was intended: as the committee of public instruction's initial report to the National Convention argued in September 1793, the *calendrier républicain* offered the world a 'new tool' with which to inscribe the 'annals of the French nation', marking each anniversary of the founding of the republic as a new year.[2] Together with reshaping the political world, the Revolutionaries endeavoured to define the republican age with a new system of days, months and years, commemorating the nation's achievements and laying the groundwork for a new future, free from the delusions of the past.

Since this inception, the republican calendar, which remained in use in France, its colonies and conquered territories from October 1793 until the end of 1805, has been understood in many ways: as the logical extension of the remapping of France's administrative geography and the reformation of weights and measures; as a statement of the Jacobins' utopian instincts and regenerative desires; as an instrument of dechristianisation; and as an

1 On the origins of the names of the months see C. von Linné and A. M. Berger, *Calendarium florae*, Upsala 1756. For example, the botanist Linneaus suggested the months of *Messis* and *Brumalis*, using similar roots as the names proposed by Fabre d'Églantine whose own adopted flowery name came from the floral games at Toulouse.

2 Also known as the *calendrier franeaise* and commonly referred to as the 'Revolutionary Calendar' in the Anglophone world. I have opted for the 'republican calendar' as a more exact translation, one that is used by the *Oxford English dictionary*, and one that emphasises the importance placed on the republic and the purported universalism of the new calendar. Other names would be attached to the calendar, such as Romme's Calendar (Carlyle), and the *Calendrier des tyrans* after Thermidor; this label was also used for the Gregorian Calendar.

analogue to the Terror, calculated, as Thomas Carlyle later noted, 'for the Gospel of Jean-Jacques'.[3] The scope of the reform was far-reaching, and certainly speaks of the grand hopes of many revolutionaries. As the primary creator of the calendar, the politician Charles Gilbert Romme, argued, France had 'opened a new book of history'; a claim that points to a particularly novel quality of the French Revolution, and one that has been taken up by subsequent revolutionary movements.[4] After 1789, it seemed possible to believe that a new era in human history could begin; it was a hope that was replayed in future revolutions, which were also accompanied by real or planned calendar reform.[5] Yet reshaping the world could be mundane. For the official at his desk interpreting the new republican laws, the farmer planning his trip to the market, the mill worker planning her work, the correspondent with her pen and paper, or, possibly more problematically, the cleric and his flock, the reform of the calendar also presented practical considerations. The great ideals of the Revolution and the practicalities of everyday life met in the structure of the new calendar. It too provides an example of how the ideals of a new political system were put into practice, an examination of the Revolution in everyday life.[6]

As such, this book offers a history of the republican calendar, and asks whether it is possible to recapture what it meant to people, even whether it meant the beginning of what we now refer to as modernity.[7] Introduced in the autumn of 1793 as the Jacobins ramped up their radical legislative programme – decreeing the general maximum on 29 September, enforcing codes of revolutionary dress and fiercely suppressing federal revolts – the republican calendar replaced the existing Catholic calendar, with a method

3 T. Carlyle, 'Terror', in *The French Revolution*, iii, London 1837, 336.

4 *Procès-verbaux du comité d'instruction publique de la Convention nationale*, ed. M. J. Guillaume, Paris 1891–1957, i. 441; M. Perrie, D. C. B. Lieven, and R. G. Suny (eds), *Cambridge history of Russia*, Cambridge 2006, 156; D. Peris, *Storming the heavens: the league of the Soviet godless*, Ithaca 1998, 89.

5 The concept of a 'new era' dominates, for example, [A. Alison], 'Fall of the throne of the barricades', *Blackwood's Edinburgh Magazine* lxiii (1848), 1393–418, and F. Bensimon, 'L'Écho de la Révolution Franeaise dans la Grande-Bretagne du XIXe siècle, 1815–1870', *AHRF* cccxlii (2005), 211–37.

6 On bureaucratic space see R. Kingston, 'The bricks and mortar of revolutionary administration', *FH* xx (2006), 405–23.

7 Recent attempts to relate 'modernity' to the Revolution include F. Fehér (ed.), *The French Revolution and the birth of modernity*, Berkeley 1990, and E. Sagan, *Citizens and cannibals: the French Revolution, the struggle for modernity, and the origins of ideological terror*, Lanham, MD 2001. James Livesey offers an historical account that attempts to avoid some of the teleology in such accounts: *Making democracy in the French Revolution*, Cambridge, MA 2001.

arguably more suitable for the new republican era. According to many Enlightenment and revolutionary texts, the existing Gregorian Calendar that had governed France since 1582 belonged to the past, a period described by the abbé Sièyès as a dark 'night of ferocity and barbarity'.[8] The old calendar had been passed into French law by Henry III, a monarch remembered by the *philosophes* for his role, as the duke of Anjou, in the St Bartholomew's Day Massacre.[9] The Revolution abolished this 'era of slavery' and retrospectively declared September 1792 to be 'Year I of the republic'. Under this new system, the year began on the observation of the autumnal equinox in Paris, which happily coincided with the anniversary of the declaration of the republic. The year continued to be divided into twelve months, but these equal months were subdivided into three decades of ten days apiece, and the remaining five days were gathered as a series of festivals at the end of the year, somewhat in the manner of the ancient Athenian Calendar. A 'Festival of the Revolution' was added every four years as an intercalary measure (that is, a leap-year). The months' nomenclature reflected the cycle of the agricultural year, from the month of wind (*vendémiaire* – September/October) to the month of fruit (*fructidor* – August/September). Days of the week were named from *primidi* to *décadi*, the tenth day being reserved for the weekly *fête décadaire* – a decadary festival that combined the reform of time with that of administration, and envisioned a weekly gathering in each new commune for fraternal celebration, morally-improving speeches and the announcement of new laws.

According to the official body that designed the new system, the committee of public instruction, the calendar offered numerous benefits to France's citizens. First, it claimed to be based on reason and the natural world, offering them a calendar derived from the movements of the sun and, eventually, linked to the geographical measure of the earth, which had been devised by leading men of science.[10] Ten-day weeks also promised a more convenient way of calculating dates, and the erasure of a surfeit of feast days would help to make France's agriculture and citizenry more productive, a cherished goal of the Enlightenment and those influenced by physiocratic economic thought. Furthermore, the calendar was based on the decimal system. As such, the hours of the day would be decimalised along with measures of money, mass, volume and length, offering a simple, unified, rationally pleasing system of

8 E. Kennedy, *A cultural history of the French Revolution*, New Haven 1989, 165.

9 Pope Gregory XIII issued a papal bull reforming the Julian Calendar on 24 February; it was enacted in France on 9 Dec. 1582, with the following day Monday dated 20 Dec.

10 The early reports suggested that the decimal second would be derived from the swing of a metre-long pendulum; the new second would also correspond more closely to a heartbeat.

measurement for those involved in commerce and manufacturing. The new calendar rejected what it termed 'superstition' and instead offered a universal measure of time that claimed to be politically neutral and which also commemorated the creation of the republic. Its inventors believed that they were placing the political and the natural world in harmony, making use of the best scientific knowledge. As Jules Michelet rhapsodised in the mid-nineteenth century, the Revolutionaries provided 'for the first time in the world ... the true measure of time'.[11] The following pages outline how a true history of this measure might be written.

Historiography

The ideological and religious aspects of the calendar, rather than its practical consequences, have dominated the ways in which scholars have written its history. For example, Alphonse Aulard put a particular emphasis on the calendar as one of the central facts in the separation between Church and State.[12] Similarly, Marxist historians, including Georges Lefebvre and Michel Vovelle, have seen the calendar as primarily an index of secularisation, linked to preceding eighteenth-century antecedents, including free-thought and decline in popular religious observance.[13] For the political right, the calendar demonstrated the godlessness of the Revolutionaries; to many, it was a handmaiden for the worst atrocities of the dechristianisation movement. In the words of the Catholic historian Jules Sauzay, since the Gregorian Calendar and Sunday observance offered a framework for the Catholic faith, the Convention 'lost no time in making these odious vestiges of the forbidden religion disappear'.[14] As Simon Schama notes, the reform provided 'an opportunity to detach republicans from the superstitions they thought embodied in the Gregorian Calendar'.[15] The new calendar became a tool for the suppression of religious almanacs, public celebration of the mass and dominical observance. These activities are taken by many historians as clear evidence of the republic's anti-clericalism.

11 J. Michelet and G. Wright, *History of the French Revolution*, Chicago 1967.

12 F. A. Aulard, *The French Revolution: a political history, 1789–1804*, trans. B. Miall, iii, London 1910, 158.

13 G. Lefebvre, *La Revolution Francaise*, Paris 1951, 218; M. Vovelle, *Religion et révolution: la déchristianization de l'an II*, Paris 1976.

14 J. Sauzay, *Histoire de la persécution révolutionnaire dans le departement du Doubs, de 1789 à 1801 d'apres les documents originaux inédits*, Besaneon 1867; L. Sciout, *Histoire de la constitution civile du clergé, 1790–1801*, Paris 1872.

15 S. Schama, *Citizens: a chronicle of the French Revolution*, New York 1989, 771.

Yet the calendar can also be seen as a positive innovation. As Patrice Higonnet notes, the calendar was intended to be a 'structuring element in the French republican's universalist endeavour to reshape space, time, citizen, and nation'.[16] For Francois Furet and Mona Ozouf, the calendar is a distillation of revolutionary ideology, an idealised conception of a republican utopia, offering a hermetic vision of the world.[17] This view is emphasised in Bronislaw Baczko's examination of the calendar's place in France's collective memory and Ozouf's definition of the calendar in the *Critical dictionary of the French Revolution*, both of which explore the differences between republican hopes and the reality of reform.[18] In these influential, philosophically-informed readings, the calendar is depicted as a contested site between a relatively artificial, prescriptive and often unpopular idea of regenerative, secular republicanism and a traditional, Catholic, counter-revolutionary world represented by adherence to the routines of the Gregorian Calendar.[19] The republican calendar is presented as an outgrowth of revolutionary rhetoric, a utopian dream without any real practical benefit.

This book also moves beyond views of the calendar as purely a dechristianising tool or as a container for republican rhetoric and symbols. Since the debates over the origins of the Revolution that climaxed during the 1989 bicentenary, the older questions about causation and class have been replaced by a greater concern with cultural matters, political rhetoric and the development and nature of the public sphere. The historiography of the Revolution has also become more varied, both in terms of subject matter and disciplinary approach.[20] All these concerns offer scope for a reinterpretation

16 P. Higonnet, 'Reviewed works: Der Französische Revolutionskalender, 1792–1805: Planung, durchführung und Scheitern einer politischen Zeitrechnung, by Michael Meinzer', *AHR* xcviii (1993), 1265.

17 F. Furet, *Revolutionary France, 1770–1880*, Oxford 1992, 141; S. Perovic, 'Untameable time', upubl. PhD diss. Berkeley 2004.

18 M. Ozouf, 'Revolutionary Calendar', in F. Furet and M. Ozouf (eds), *A critical dictionary of the French Revolution*, Cambridge, MA 1989, 538–47; B. Baczko, 'Le Calendrier républicain', in P. Nora (ed.), *Les Lieux de mémoire*, Paris 1984, 37–83; M. Halbwachs, *On collective memory*, trans. L. A. Coser, Chicago 1992. Baczko also explores the utopian aspect of the Revolutionaries and Ozouf's work on festivals expands on this view of the calendar: M. Ozouf, *Festivals and the French Revolution*, Cambridge, MA 1988. For a further critique see J. F. Byrnes, *Catholic and French forever: religious and national identity in modern France*, University Park, PA 2005, 47.

19 For an account of how republicanism and Catholicism could be amalgamated see S. Desan, *Reclaiming the sacred: lay religion and popular politics in revolutionary France*, Ithaca 1990.

20 It is perhaps telling that the two most recent studies that touch on the republican calendar have emerged from the disciplines of literary studies and intellectual history:

of the calendar. Furthermore, time itself has also become the subject of historical study. For example, David Landes and Gerhard Dohrn-van Rossum have produced large-scale histories of the development of clocks and calendars in the western world, while other studies have begun to examine how time has been understood at different points and in different places in the past.[21] Time has been shown to be more malleable, subjective and multi-levelled than previous accounts have assumed and intimately associated with processes of self-fashioning and identity. The growth in the study of the history of collective memory and commemoration has also demonstrated the possibilities of studying the past in the present.[22] The claims that the calendar made about the development of the French nation provide further grounds for revisiting it; if Rebecca Spang is correct and the 'revolution in the study of the Revolution has left a much grander historical narrative, about the characteristics and chronology of "modern" life, largely untouched', then the foundations of this narrative of modernity ought to be revisited.[23] Was the calendar an attempt to provide a narrative of origins, or was some other paradigm in operation? What did it say about the process of change? And what did this mean to the men, women and children who lived through the Revolution? Did it provide an opportunity to understand or redefine oneself in time, and in relation to others in the community or nation?[24] The calendar offers a path to follow in providing answers to some of these questions, forming a point of contact between the regenerative aims of the Revolutionaries and the everyday practices of timekeeping.

Rebecca Spang also draws attention to the influence, whether implicit or explicit, of psychological questions and assumptions at work during the Revo-

Perovic, 'Untameable time', and W. Nelson, 'The weapon of time: constructing the future in France, 1750 to Year I', unpubl. PhD diss, Los Angeles 2006.

21 D. S. Landes, *Revolution in time: clocks and the making of the modern world*, Cambridge, MA 1983; G. Dohrn-van Rossum, *History of the hour: clocks and modern temporal orders*, Chicago 1996; R. Poole, *Time's alteration: calendar reform in early modern England*, London 1998; F. Maiello, *Storia del calendario: la misurazione del tempo, 1450–1800*, Turin 1994; R. Beck, *Histoire du dimanche: de 1700 à nos jours*, Paris 1997.

22 Important studies include S. J. Sherman, *Telling time: clocks, diaries, and English diurnal form, 1660–1785*, Chicago 1996, and H.-J. Voth, *Time and work in England, 1750–1830*, Oxford 2000. The history of memory can also be approached via Halbwachs, *Collective memory*, and P. Nora, *Realms of memory: rethinking the French past*, New York 1996. See also N. Elias, *Time: an essay*, Oxford 1992, and N. D. Munn, 'The cultural anthropology of time: a critical essay', *Annual Review of Anthropology* xxi (1992), 93–123.

23 R. L. Spang, 'Paradigms and paranoia: how modern is the French Revolution?', *AHR* cviii (2003), 119–47.

24 B. R. Anderson, *Imagined communities: reflections on the origin and spread of nationalism*, London 1983.

lutionary period and to the materiality of the past, notably the 'discursivity of physical things and the physicality of discourse'.[25] Ken Alder has shown how physical objects (in particular, the fabrication of muskets) might, as it were, be inscribed with ideology, and in turn such objects informed thinking about what the Revolution and the republic might be, while Spang also suggests that historians have resorted to a language of madness, or Jacobin 'instinct' to explain the violence and Terror of the 1790s.[26] These insights must surely apply to the republican calendar. It was an artefact as well as a symbol, leading to the creation and sale of almanacs, engravings, clocks, watches, sundials, songs, poems and calendars, whose construction drew on a certain republican ideology and scientific belief. The calendar, and its derivatives, attempted to inscribe a concept of the Revolution on the individual, forming the structure of their everyday life, timing the events of birth, life and death, commercial contracts and moments of leisure and, through exposure to a series of virtuous or useful concepts and objects, was designed to lead to their civic regeneration.[27] It was, as Romme recognised in the language of his day, a 'psychological' tool – a material, printed object and a structure for physical, bodily practice, organising everyday life in a time of turbulent, revolutionary social and political change. His calendar replaced saints' and feast days with agricultural tools and plants or with festivals dedicated to 'Work' or 'Public Opinion'.

Despite the calendar's wider implications, it could be argued that it has not received the attention it deserves, with only two historical studies produced between World War II and the Millennium, James Friguglietti's unpublished doctoral dissertation (1966) and Michael Meinzer's study (1992) on the calendar's adoption in Marseilles and Provencal villages.[28] Moreover, these

25 Spang, 'Paradigms and paranoia'.

26 K. Alder, *Engineering the revolution: arms and enlightenment in France, 1763–1815*, Princeton, NJ 1997. For a positivist analysis of the relationship between ideas and objects see S. L. Macey, *Clocks and the cosmos: time in western life and thought*, Hamden, CN 1980.

27 The vital links between measurement and materiality are underlined by Witold Kula: 'measure is intimately connected with man and the things he values above all others: land, food and drink': *Measures and men*, Princeton, NJ 1986, 17.

28 J. Friguglietti, 'The social and religious consequences of the French Revolutionary Calendar', unpubl. Phd diss, Cambridge, MA. 1966, and 'Gilbert Romme and the making of the French republican calendar', in D. G. Troyansky, A. Cismaru and N. Andrews, Jr (eds), *The French Revolution in culture and society*, Westport, CN 1991, 13–22; Meinzer, *Der französische Revolutionskalender*. See also G. G. Andrews, 'Making the Revolutionary Calendar', *AHR* xxxvi (1931), 515–32; Baczko, 'Calendrier républicain'; S. Bianchi, 'La Bataille du calendrier ou le décadi contre le dimanche: nouvelles approches pour la reception du calendrier republicain en milieu rural', *AHRF* cccxii (1998), 245–64; and M. Shaw, 'Time and the French Revolution, 1789–Year XIV', unpubl. PhD diss., York 2000, and

view the calendar less in terms of its wider cultural context than through the lens of the clash between religion and republicanism.[29] The two historians disagree about the extent of the calendar's success, but measure that success in much the same way. Both concentrate upon its social and political history, particularly its relationship to dechristianisation, and both emphasise its political importance and its local responses. Friguglietti, whose sources are in a large part derived from areas with a strong Catholic identity, particularly the Department of the Nord which was troubled by counter-revolution, concludes that 'in the eyes of most Frenchmen, the new calendar represented an inconvenience, an absurdity that served no practical purpose', while Meinzer finds more evidence for its observance in Marseilles, for example amongst pro-Jacobin fishing fraternities which took the calendar's ten-day week as the basis for their meetings.[30] As well as pointing to the geographical and religious differences that underpinned the reception of the calendar, the two studies also show how different sources can lead to different interpretations. For example, Friguglietti's account relies largely on administrative and judicial reports, which highlight infractions to the calendar and, hence, its unpopularity, while Meinzer's use of printed materials and quantitative evidence of meeting times emphasises the extent to which the calendar informed official (and pro-Jacobin) behaviour. Even though these two studies disagree on the practical effects of the calendar, they largely see it in terms of success or failure and as a clash between republicanism (or, perhaps, Jacobinism) and a reluctant, religious or counter-revolutionary population. Baczko and Ozouf also emphasise this distinction, proposing a binary opposition between Revolution and counter-revolution, between secular society and religion, and between the 'real' and the 'imaginary'.[31] While there is much evidence for such a clash, I suggest that while it was intended to sweep away existing modes of behaviour, the calendar could be accommodated by many people in their daily life, in a similar manner to the responses of the citizens of the Yonne to the republic's religious policy, where, as Suzanne

'Reactions to the French republican calendar', *FH* xv (2001), 4–25. The calendar is considered as part of wider, sociological studies in E. Zerubavel, 'The French republican calendar: a case study in the sociology of time', *American Sociological Review* xlii (1977), 868–77, and *Hidden rhythms: schedules and calendars in social life*, Chicago 1981.

29 Perovic, 'Untameable time'; Nelson, 'The weapon of time'. Noah Shusterman's work on the calendar appeared too late to be considered in full here: *Religion and the politics of time: holidays in France from Louis XIV through Napoleon*, Washington, DC 2010.

30 Friguglietti, 'Gilbert Romme', 158. Meinzer's study makes extensive use of the records of the dates of meetings: *Der französische Revolutionskalender*.

31 B. Baczko, *Lumières de l'utopie*, Paris 1978, 218.

Desan has shown, people could combine republican sentiment with an attachment to their local church.[32]

In this way, the calendar may be seen as something of a referendum on the Revolution as a whole, with its use or opposition potentially mapping support for the republic. Yet this view needs to be scrutinised, not least because opposition could mean many things (and 'the Revolution' was by no means a single, totalised event). In particular, research on the Vendée, Federalism and counter-revolution has revealed the many layers of revolt, ranging from principled religious and ideological opposition to what might be seen as unchanging local hostility to external intervention.[33] The methods of anthropology and sociology have been put to use to explain the complexity of provincial or regional politics and reactions to revolutionary events and processes.[34] Furthermore, a case can be made for the impact and durability of the calendar in private and public life. As Carlyle also commented, 'the Romme Calendar, in so many Newspapers, Memoirs, Public Acts, has stamped itself deep into that section of Time: a New Era that lasts some Twelve years and odd is not to be despised'.[35] Meinzer's research has revealed the contours of some of these impressions, notably among the Marseilles fishing fraternity, which appeared to have adopted the patterns of the new calendar, particularly during the Consulate. By examining a series of departmental records and reports received in Paris from across France, the account offered in this book also suggests that the calendar's consequences, and the extent of governmental support for its introduction, may have been underestimated by historians, particularly as understanding of how time and calendars were seen in the past remains something of an under-explored area. Friguglietti's assessment of the calendar as an absurd failure avoids a serious engagement with the areas of French life in which the calendar can be seen to have made a significant impression.

The aim of this volume is not simply to measure the success or failure of the calendar, but to use it to investigate the ways in which reform was perceived and understood by contemporaries. As such it attempts something broader: a 'history of time' during the last decade of the eighteenth century. Although existing studies have examined certain aspects of the calendar in

32 Desan, *Reclaiming the sacred.*

33 For a synthesis of studies on the relationship between local and national politics during the Revolution see A. I. Forrest, *Paris, the provinces and the French Revolution*, London–New York 2004.

34 C. Tilly, *The Vendee*, Cambridge, MA 1964; C. Lucas, *The structure of the Terror: the example of Javogues and the Loire*, Oxford 1973.

35 Carlyle, *French Revolution*, ii. 295.

detail (notably, the 'high' politics of its introduction by the Convention as recorded in the published records of the committee of public instruction, and by examining a certain number of departmental responses), they have not placed the reform within its wider cultural context. The introduction of the calendar needs to be assessed alongside the other local policies of the Revolutionaries, not just republican attitudes towards dechristianisation, an aspect that has come to be closely linked to the calendar in later accounts of the Revolution, often reducing the calendar to a battle between the Catholic *dimanche* and the Jacobin *décadi*. Furthermore, while in many ways Friguglietti broke new ground with his detailed examination of the calendar, our understanding of the importance of cultural politics has greatly expanded, suggesting that cultural innovations such as the calendar or changing codes of dress or social etiquette were not peripheral to the Revolutionary experience but were essential to the enunciation of politics in the 1790s. Everyday behaviour is now seen as potentially more political than it once was.[36] Finally, and most important, in order to understand its relationship to these social and cultural codes, the calendar needs to be related to the broader context of eighteenth-century temporality. What, for example, did it mean to reform the structures of time in the 1790s? Clearly, it is anachronistic to compare modern time consciousness and often highly-timetabled twenty-first-century lives, and their pervasive 'language of busyness', with modes of eighteenth-century behaviour. What, in contrast, did time mean in the eighteenth century, how mutable was it, and to what extent did reform disrupt people's lives?

Sources and scope

This book presents a more nuanced interpretation of the calendar, one which is based on the premise that the calendar can only be properly understood and assessed in the light of this cultural and temporal context. Seen in this way, the calendar can be shown to have heralded significant consequences not just for everyday life in France and its sister republics but also for the history of timekeeping more generally. The reorganisation of the days of the week and a concern with citizens' time can be viewed in terms of centralisation and placed within the context of a longer-term extension of official power, a concern with legislation and regulation that extended back into the *ancien*

36 L. A. Hunt, *Politics, culture, and class in the French Revolution*, Berkeley 1984; K. M. Baker (ed.), *The French Revolution and the creation of modern political culture*, i, Oxford 1987.

régime and forwards into the nineteenth century. Public order, religion and work all received the regulatory attentions of civil and religious authorities, restricting the times when certain activities, such as drinking or dancing, could take place, or requiring attendance at particular rituals. In a similar fashion, republicans had their own concerns with these matters, not least because riot or dissent were seen as revealing counter-revolutionary shadows. Religion also became one of the fault lines of French society, with a complex, and often violent, debate about toleration and loyalty, and the new calendar became one of the implements that forced the two sides apart. Finally, the Jacobin republican project was also an economic one, concerned with both equality and production. Shipyards were reorganised, material marshalled, and manufactories of all kinds, from arms factories to watchmaking, promoted. The calendar formed an important plank in such economic policy, or at least offered a means of imagining a productive nation, negotiating the aesthetics of the countryside and town, agriculture and manufacturing. Watches and clocks became symbols of modernity.

Timepieces – watches, clocks and sundials manufactured in a great range of styles and levels of complexity – fulfilled a practical function, but their cultural importance, or symbolism, may be seen on a more metaphorical level. The measurement of time was not metaphysically neutral but could make a number of symbolic connections. Primarily, it is a metaphysical enterprise designed to co-ordinate 'the earthly organisation of man's life with the temporal order of the universe'.[37] Since the medieval period, and indeed before, the clock has been a symbolic instrument and illustrated the regularity, order and harmony of the world. The Enlightened mind seized on the physical and mechanical attributes of clockwork and its representation of order and rationality.[38] The calendar represented a continuum of these temporal metaphorical possibilities. Although the clock and the calendar measured time on different scales, they were closely linked in the eighteenth-century mind. Other aspects of time could be a cultural resource. Personal 'memory maps' link the events of individuals' lives with historical occasions or memorable days in the calendar. Indeed, as Tim Ingold argues, the very landscape and its cultural associations are suffused with temporality. The seasons and agriculture are constant reminders of the importance and rhythms of time, while memorials, graves and dates placed on buildings and

37 J. T. Fraser, *Time as conflict: a scientific and humanistic study*, Basle–Stuttgart 1978, 131.

38 J. W. Konvitz, *The urban millennium: the city-building process from the early Middle Ages to the present*, Carbondale–Edwardsville 1985, 18–19; C. Menéres, 'L'Espace du temps: l'horloge', *Temps libre* ii (1981), 9–17; O. Mayr, *Authority, liberty and automatic machinery in early modern Europe*, Baltimore 1986, 77, 119.

statues present other stories about time.[39] The time of the self, of the community and the nation all have separate tempos and importance, but also relate to one another. The introduction of the republican calendar affected all these levels of identity. A contemporary noted: 'You know, CITIZENS, the importance of a good division of time [...] it links everything that is done today, with those generations that are to come & even with foreign nations.'[40] Time was a matter of great significance.

The potential scope for studying such an elusive topic as time, not to say the French Revolution, is, then, huge. Thankfully, the calendar's close links to the bureaucratic state offer a means of structuring such an analysis, as well as ensuring that many of the deliberations that informed the calendar's introduction are recorded relatively systematically in official archives. The papers of the central body responsible for the calendar, the committee of public instruction, can be consulted at the Archives nationales in Paris, and many of their minutes were edited and published in the nineteenth century.[41] Other departments at a national and Departmental level provide evidence for the calendar's introduction; these political and administrative records have been used by Friguglietti and Meinzer in their social and political accounts of the first years of the calendar. Police and municipal records offer a view on the problems created by the calendar, not least evidence for those who opposed its introduction. Friguglietti examined a series of records made by the Paris authorities that are now held by the Archives nationales, and I have drawn on these fonds along with several other series, particularly those from the Directorial period.[42] These files help to track the ways in which administrative deliberations placed the calendar at the heart of the new French state. Departmental case studies allow a picture to be drawn of the ways in which the calendar was implemented locally. Existing studies have been supplemented by research in the Departmental and municipal or city archives of the Vaucluse (Avignon), the Doubs (Besaneon), the Côte d'Or (Dijon), the Loire-Inferiere (Nantes; now the Loire-Atlantique), the Somme (Amiens) and the Nord (Lille). Records of temporal regulation before and after the Revolution have been examined as well as the contents of the Isère Depart-

39 Tim Ingold argues that time and the temporality of the landscape provides a point of contact between archaeology and anthropology and views time on a variety of levels: 'The temporality of the landscape', *World Archaeology* xxv (1993), 152–72.

40 Gugoulin, Apt, to the comité d'instruction publique, 28 thermidor II (15 Aug. 1794), AN, F^{17}1355.

41 *Procès-verbaux*.

42 The following fonds at the Archives nationales, Paris, are particularly useful: DIII, Comité de legislation; $F^{1}C^{III}$, Esprit public et elections; and F^{17}, Instruction publique.

mental archives reproduced in the *French Revolution Research Collection*.[43] The records tend to show a note in the minutes adopting the calendar, followed by distribution of manuals and posters announcing the new calendar. Sometimes the records reveal discussions about how the calendar might be implemented; more often there are records of speeches exhorting the local community to follow the new calendar. Later records, particularly those related to local policing and the courts of the *juges de paix*, detail the fines and imprisonment imposed on those caught working on the *décadi*. While such records remain largely silent on whether the general public used the calendar, they speak much more loudly of a growing administrative machine that quickly came to terms with the new calendar. Serge Bianchi, for example, notes the speed with which communal authorities implemented the calendar and its system of festivals in several villages in the Côte d'Or. Jack Thomas shows how the reform of markets in the south-west bridged the gap between everyday life and republican hopes for reform.[44] I have also worked through the nationwide reports on Departmental '*esprit public*' held in the Archives nationales, which are obviously even more mediated and concise than records maintained locally, but are still useful for providing a picture of the calendar; not least because the *fêtes décadaires* and the reorganisation of fairs and markets became one of the questions to which the department had to respond regularly.[45]

In this way, an attempt has been made to synthesise existing studies and new research, offering an overview of reactions to the calendar across France, and examining several regions in more depth. This approach extends those surveyed by earlier works to draw out some of the factors involved in responses to the calendar across France. To what extent, for example, do geographical or social differences help to explain reactions to the calendar? Zones of counter-revolution are encompassed in the west by the Loire-Inférieure, in the Midi by the Vaucluse, and by the mountainous, wooded regions of the Doubs in the east of France. These regions contained towns with a strong Jacobin base (such as Nantes and Besançon) from where concerted campaigns to enforce the calendar would emerge. Other parts of the Doubs and its neighbour, the Côte-d'Or, provide examples of somewhat more politically neutral areas. As a result of a scheme to re-establish Swiss

43 These reproduce the dossiers on the republican calendar held in the Isère departmental archives. I have also consulted the full series of printed inventories of departmental archives, many of which include information about the nature of administrative concern with the calendar.

44 Bianchi, 'La Bataille du calendrier'.

45 See the bibliography for a fuller list of these reports.

migrants, the Doub's capital, Besaneon, became the centre for French watch-making, providing information on both the material culture of clocks and the relationship between proto-industrial production methods and timekeeping; as part of the Franche-Compté, the Doubs was also formed from an area with a strong sense of local traditions and distinctiveness. Similarly, the Department of the Nord represents another liminal region, an area only recently 'made French' by the time of the Revolution and with acute religious sensibilities; the Côte-d'Or and the Loire-Inférieure were also formerly part of distinct *pays d'état*, with regional assemblies.

The relationship between Paris and the provinces provides another important way of looking at the calendar, revealing the tensions involved in the centralising processes that formed an inevitable part of the creation of a republican bureaucracy. Time-control in Paris, the epicentre of administrative concern with public behaviour both before and after 1789, has therefore been traced, with police records proving particularly useful in charting street-level responses.[46] Other archives, such as that of the Bureau des longitudes and Charles Gilbert Romme's papers in Milan and the Bibliothèque nationale, have also been consulted for further details of the calendar's creation. It has also been important to look backwards, at the records of 'temporal policing' produced before the Revolution, examining records of town clocks and church bells, the regulation of markets, and the attempts to control drinking on Sundays and other feast days. From these a picture of extended local concern with time-based regulation emerges, a pattern which is taken up by the attempts to introduce the republican calendar. Finally, places in which the calendar appears more obliquely have been considered: personal papers and diaries, and, most systematically, the meetings of a series of *juges de paix*, attempting to see whether their schedules replicated that of the Gregorian or republican calendar.

What emerges is perhaps to be expected: a marked difference between town and country, and adherence following periods of central concern for the calendar. Yet, there are also signs of perhaps surprising rural observance and, especially, increased observance from the post-Fructidor Directorial period onwards (that is, following the purge of conservatives from the *Corps législatif* in *fructidor* Year V/September 1797). Like the metric system, which took several generations to implement, a change in calendar could well have taken place given enough time and political will. Finally, the sources also reveal the seriousness with which the

46 'Procès verbaux des commisaires de police, 1791–Year X', Archive de la Préfecture de Police de Paris, Paris.

calendar was often viewed, and the extent to which administrative effort was devoted to it. The calendar is revealed as a continual, unavoidable presence in the nascent administrative machinery of the republic.

Although concerned with time, this book is largely structured by theme, rather than by chronology. It begins with a general consideration of time in the eighteenth century and an overview of the history of the republican calendar, followed by chapters examining time during the revolutionary era, from the macro level of republican ideology, working its way towards the micro, from the days of the week to the attempt to decimalise the hours of the day.

1

Time and History

The varieties of temporal experience

Time is an elusive category. On the one hand, as the literary critic Stuart Sherman has noted, the eighteenth century was marked by 'its unprecedented passion for chronometric exactitude, in its timepieces and in its prose'.[1] Time could be understood as a metaphor for regularity, order and control, with a Supreme Being setting the world in motion like clockwork. Montesquieu, for example, privileged the machinery of clockwork as a simile in the 1757 foreword to his *L'Esprit des loix*: 'the spring that makes republican government move, as *honor* is the spring that makes monarchy move'.[2] Yet, as Michèle Perrot suggests, time could also be 'extendable, elastic, fluid according to the light, lengthened or shortened following the seasons, the time of the sun, nonchalant and dreamy'.[3] Crudely put, time in the eighteenth century can be depicted as either 'baroque', pre-industrial, dominated by the seasons, the sun and the irregular routines of farming, the small workshop and the patterns of the Catholic year, or as a period of transition towards what E. P. Thompson has termed 'industrial work time', albeit perhaps less to do with factories and industrialisation than linked to the aesthetic order of neo-classicism, the obsession with horology shown in the *Encyclopédie* and a desire to control and regulate on behalf of society's elite.[4] In part, such differ-

1 S. J. Sherman, 'Telling time: clocks and calendars, secrecy and self-recording in English diurnal form, 1660–1795', unpubl. PhD diss. Columbia 1990, 9.

2 C. de Secondat Montesquieu, *De l'Esprit des loix*, i, London 1757, author's preface. Rousseau argued that 'The philosopher's time-piece is evenness of temper and tranquility of mind; he is always in his own time and knows it exactly': *Emilius and Sophia, or a new system of education*, ii, London 1783, 59. See also Mayr, *Authority, liberty and automatic machinery*.

3 Sherman, 'Telling time', 9; Michèle Perrot, *Les Ouvriers en grève*, Paris 1974, 271.

4 Michel Foucault's work provides the classic statement of this impulse, but see also that of Robert Muchembled such as *Popular culture and elite culture in France, 1400–1750*, Baton Rouge 1985. These views have been stridently opposed by many, particularly in regard to

ences come from the varieties of temporal experiences that are present in any society, and reflect the different concerns of the culture of the court, the salon, the poet, the workshop or the street, as well as the contradictory nature of eighteenth-century France.[5] They are also a reflection of economic realities and the demands of labour. Time reflected and helped to reinforce the numerous divisions within society, just as much as it might unify it with a common calendar.

The sociologist Norbert Elias's reflections on time offer a way to negotiate time's relationship with power and authority. Elias stressed the multilayered nature of time and the subtle ways in which it serves to underscore social relations. For him, time is a 'social institution' that takes on different meanings in different social contexts, helping to define the social constraints or 'social habitus'.[6] In eighteenth-century France, the many levels of clock time, differing work practices and the demands of the agricultural year, together with the religious and civil annual cycle, created a dense temporal texture, which served to underline the lineaments of power. For example, *ancien régime* France could be depicted as an absolutist state, dominated by the symbols of an unchanging monarchy, and at the same time as an increasingly capitalist, consumerist and dynamic society. The courtly routine of Versailles, with the king's public rising, dining and retiring, contrasted with the bustle and rush of business at the Bourse. Ordered routine and a regal, leisured approach to the arrangement of the day helped to create the mystique of power at the centre of Versailles, from ritual awakening and dining, to the days spent on the hunt. Even the magistrates of the Paris *parlements*, so self-consciously sophisticated and modern in other ways, recognised the magisterial tone that could be set by stately procession, lengthy, etiquette-bound meals and sacred ceremonies, as their hospitality records attest.[7] Civil authorities across France, whether gilds or city councils, also saw the importance of a system of festivity and calendar ritual. As the eighteenth-

medicine and lunacy: E. P. Thompson, 'Time, work-discipline, and industrial capitalism', *Past and Present* xxxviii (1967), 56–97.

5 On the debate on the fluid or stratified nature of the '*ancien régime*' see C. Jones, *The great nation: France from Louis XV to Napoleon*, London 2003, pp. xx–xxi.

6 Elias, *Time*. See also B. Barnes, 'Between the real and the reified: the sociology of time', in S. Loyal and S. Quilley (eds), *The sociology of Norbert Elias*, Cambridge 2004, 59–74. There is an extensive literature on the relationship between time and society, particularly from an anthropological or sociological perspective: Munn, 'The cultural anthropology of time'. The history and study of time and timekeeping can be approached via Landes, *Revolution in time*; Dohrn-van Rossum, *History of the hour*; Maiello, *Storia del calendario*; and B. Adam, *Timewatch: the social analysis of time*, Cambridge 1995.

7 J. H. Shennan, *The parlement of Paris*, Ithaca 1968.

century shopkeeper and glazier Jacques-Louis Ménétra's journal reveals, the ritual year of gild festivals helped to give shape to the fluid patterns of artisanal life.[8] But the multiplicity of timeframes could also undercut and challenge the existing order. The pressures of buying and selling, and the new public spaces of the salon or coffee shop created another kind of temporality. On the street, the bustle of modern life was impressing itself on journalists such as Louis-Sebastien Mércier, whose account of 'How the Day Goes' in the *Tableau de Paris* (written after the Revolution) presents the different hours of the day as 'background of a noisy and speedy whirlwind, calm, or of movement'.[9] The day begins with farmers and vegetable sellers heading into the city, the environs of which then reflected the movements and timings of different social groups, from lawyers to speculators heading to the Bourse and, late at night, the rich in their carriages speeding through the streets.[10] Such journalistic generalisations point to the ways in which haste and leisure, the routine of the hours of the day or a sense of increasing hustle and bustle could imbue time with certain moral qualities: how time was used, whether leisure signified power, idleness or waste. It also served to create networks of community and hierarchy: whether through hours taken in the companionship of a drink in a bar between jobs at the workshop, or among the elite at the card table.[11]

As well as power, time could signify difference. Town and country can be seen as different temporal zones, with different technologies and patterns of work and socialising creating separate senses of time. Similarly, social classes were separated by their access not only to economic freedom from work, but also to the simple technologies of the candle or oil lamp, let alone a clock or watch. Time could also differentiate between young and old, between men and women, between secular and sacred, between French and other. Travellers regularly commented on different timekeeping practices, such as the Italian habit of ringing the start of the new day at sunset, rather than sunrise or midnight (the new day started at midday on naval ships).[12] For the men and women of the eighteenth century, intrigued by past and exotic societies, time was never a settled matter. As the entry on 'Tems' in the *Encyclopédie* concludes, 'the measures of time are arbitrary & may vary among different

8 R. A. Schneider, *The ceremonial city: Toulouse observed, 1738–1780*, Princeton, NJ 1995.

9 L.-S. Mercier, *Le Tableau de Paris*, ed. J. Kaplow, Paris 1979, i. 146.

10 D. Garrioch, *The making of revolutionary Paris*, Berkeley 2002, 15–26.

11 On the hours of the day in the city see also ibid. For an English comparison on the economics of leisure time see Voth, *Time and work*.

12 For example, J. G. Keysler, *Travels through Germany, Bohemia, Hungary, Switzerland, and Lorrain*, London 1756, i. 301, and pp. 100, 251, 280.

peoples; only the moment is universal'.[13] The attempts by anthropologists in the twentieth century to characterise the ethnology of time had their roots in the concerns of the Enlightenment. Just as the clock metaphor had infiltrated philosophical and theological rhetoric, interest in temporality informed discussion of society, whether at home or abroad.

The existing Gregorian Calendar, perhaps in part due to a general interest in the exotic, classical or oriental, was not seen as necessarily natural, or indeed fixed, at least by the elite. Any educated French man or woman knew of the variety of calendars which existed, not only in Europe (Britain had used the Julian Calendar until 1752), but also amongst different religious groups. A reader of the *Encyclopédie* who had reached the volume considering the third letter of the alphabet would also be aware of the numerous calendars used throughout history, from the Egyptians and the Greeks to the reforms of Julius Caesar and Gregory XIII's *Inter gravissimas* of 1582. Time, as the dominance of the clock metaphor in philosophical discourse reveals, had infiltrated the language and thinking of the Enlightenment.[14] Thinking about the calendar was no different, a development that can be followed in the pages of the *Journal des sçavans*, founded in 1665 by Denis de Sallo, a member of the Paris *parlement*, published fairly regularly until the 1790s, and revived as the *Journal des savants* in 1816.[15] During this period it published a range of articles and reviews that reveal three phases of concern about the calendar and time. The first of these, which dates from the last quarter of the seventeenth century to the early years of the eighteenth, reveals the religious and political dimensions of the scientific study of the calendar. Establishment figures, such as the royal astronomer Cassini, who had been 'lent' to the French king by the pope, debated the future accuracy of the equinox within the context of the Gregorian Calendar in the face of opposition from the reformed calendar and proposals by men such as Touraine, who spent forty years refining and defending his *Calendrier perpetuel et universel*. (Unfortunately, it was neither perpetual nor universal.) Here, the authority of the papal reforms is under question; defence comes not from the papacy, but from the officials of the state, such as Cassini and others who strove to defend the accuracy of the Gregorian Calendar. In 1708 H. Klaufing argued in favour of the Protestant reformed calendar, and Samuel Krob proposed a new ecclesiastical calendar in 1709. Religion, however, began to fade in this debate from the 1700s.

13 'Tems', in D. Diderot, *Encyclopédie, ou, dictionnaire raisonné des sciences, des arts et des métiers, par une société des gens de lettres: mis en ordre & publié par* M. *Diderot* ... *& quant à la partie mathématique par* M. *d'Alembert*, Geneva 1772, xvi. 93–6.

14 Macey, *Clocks and the cosmos*, 62, 70–2; Mayr, *Authority, liberty and automatic machinery*.

15 S. H. Steinberg, *Five hundred years of printing*, Harmondsworth 1955.

What followed was a rediscovery of ancient calendars, in particular, those of Rome and Egypt, and the publication of interpretative aids and concordances. Calendars were published to help those studying the past, but also to enable modern dates to be put in the classic form, should the reader wish. The final shift was to an increased concern with the past and its divisions and with the history of the calendar. For example, in 1740 an account of Gregory XIII included a relatively detailed history of the reform of the Gregorian Calendar, while in 1743 the Egyptian calendar was printed, and in 1746 the Calendar of Naples came under discussion. A new genre, the 'calendrier historique' also begins to be published, and regularly reviewed in journals from the mid-eighteenth century. Furthermore, the usefulness of the calendar had been reassessed. It was not simply used for the dating of religious festivals, or for astronomic science, but was a daily tool. This had long been the case for some, but by the mid-eighteenth century, a 'calendrier journalier', as one publisher advertised it in 1746, was useful not just for the historian or chronologists, but many men of business, law and the professions. A *Calendrier des dames*, which included not only a list of kings, but a female portrait to accompany each month, was published in 1749.

National calendars differed as well as the hours of the day. Frederick II had only relatively recently adopted the Gregorian Calendar in Prussia (1775), and the political satires generated by the shift in Britain in 1752 may have remained in people's memories in 1789.[16] Russia still followed the Julian Calendar, and travellers reported on the different practices of inhabitants of other countries such as Iceland. For example, one visitor noted that the men and women of Iceland only observed two seasons, and had eight 'hours' during the day, which they could determine by the sun and sea. He also noted that 'watches are very rare among them; every peasant, however, has an hour-glass'. Such things were often the material for the front matter of French almanacs and learned journals, as well as being given substantial space in the *Encyclopédie*, which also pointed out several defects in the Gregorian Calendar.[17] The influence of environmentalist thought also encouraged a belief that geography determined the approaches to time of different cultures. Religion also figured strongly in these temporal differentiations, with the Jewish or Islamic Calendars helping to form the identities of these groups; Protestant and Catholic were divided and defined by their fondness for or dislike of saints days or particular religious festivals.

16 Poole, *Time's alteration*.

17 U. Van Troil, 'Letters of Iceland', in J. Pinkerton (ed.), *Voyages and travels*, i, London 1808, 657; 'Calendrier', *Encyclopédie*, ii. 555.

The *Almanach des honnêtes-gens*

The symbolic power of the calendar meant that interventions could be fraught with dangers. The personal dangers of reforming the calendar had been demonstrated some eighteen months before the Revolution. On 9 January 1788 Sylvain Maréchal was sentenced and imprisoned by the Paris *parlement*, and his work, the *Almanach des honnêtes-gens*, condemned as impious and blasphemous.[18] Maréchal, the son of a Parisian wine-merchant, had trained as a lawyer, but never practised and had instead looked for a career within the 'republic of letters', becoming a bookseller and avowedly atheistic author.[19] At the age of thirty-eight he published the *Almanach* that was to lead to three months' imprisonment and exile. This work was a calendar in which the days of the year recorded the births and deaths of great men of literature, philosophy, art and science. It contained many features that would resurface in the republican calendar, in particular the structure of the year. Maréchal's almanac was published 'for the current Year', which was, the *Almanach* declared, 'The first year of the reign of Reason'.[20] In his scheme, the year began in March and the months were subdivided into three 'décades' of ten days. Clearly, Maréchal was convinced of the benefits of decimalisation. Aware of the calendars of ancient Egypt and classical Greece, the remaining five days were ordained for festivals called 'épagomènes', dedicated to moral themes, such as the *fête de l'amour* proposed for 31 March and the *fête de la reconnaissance* on 31 August. Maréchal also provided alternative, Roman names for the twelve months: *Princeps*, *Alter*, *Ter*, *Quartile*, *Quintile*, *Sextile*, *Septembre*, *Octobre*, *Novembre*, *Décembre*, *Undécembre* and *Duodécembre*; this structure and the classical allusions resurfaced in the republican calendar in 1793.

Maréchal suggested that this new formulation of the calendar could be consulted by 'the Catholic and the Protestant, the Lutheran and the Anglican, the Christian and the Muslim, the Idolater, and the Hebrew' alike. Such a suggestion was provocative, even blasphemous, especially when the rest of the calendar is considered. Maréchal's publication was clearly an attack on the commemoration of saints whose days were replaced by

18 Sylvain Maréchal (15 Aug. 1750–18 Jan. 1803).

19 C. Jones, *The Longman companion to the French Revolution*, Harlow 1988, 370; F. Pouy, *Recherches sur les almanachs & calendriers*, Amiens 1874, 104–11; M. Dommanget, 'Sylvain Maréchal, précurseur du calendrier révolutionnaire', *International Review for Social History* iii (1938), 301–34.

20 S. Maréchal, *Almanach des honnêtes-gens*, *l'an premier de la raison*, Paris 1788, 100–1.

humanist heroes. One date remained empty: the 15 August, which, rather than some prescient homage to Napoleon, was Maréchal's own birthday. Christ and Newton shared the 25 December as their common *anniversaire*. More respectfully, perhaps, Maréchal's father was also included in the calendar. None the less, Maréchal was premature in declaring that the age of reason had begun. The almanac was not seen, or intended, as just a work of whimsy or a piece of entertainment, but was taken as an attack on the Catholic religion and was denounced as such by the *parlement de Paris*. Although essentially a calendar, it was also called an *almanach*, a word not without suggestions of occult prognostication or, in this case, vanity (as well as filial devotion). Maréchal's atheistic reputation ensured that the case was prosecuted and the work condemned as a 'work of impiety, of atheism and of folly'. All remaining copies of the almanac were impounded and burned, while its printer, Charles Gailleau, found himself exiled three leagues from the city and fines were to be imposed on anyone who attempted to sell the condemned work. None the less, Maréchal remained undaunted by these punishments and swiftly began to publish similar material after his release.

Despite Maréchal's problems, this range of temporalities suggests that introducing a new calendar was not necessarily a quixotic aim. Both the calendar and clock time were undergoing a process of transformation. Anthropologists have tended to see time as a key category of human society and, as such, a relatively unchanging feature and an aspect of the everyday particularly resistant to bureaucratic change. But, as Elias theorised, time is also historical and tied to structures of power. Its patterns do change and offer the opportunity to challenge power itself. Jacques Le Goff has argued for the growth of 'merchant time' at the expense of 'clerical time'.[21] Landes and Dohrn-van Rossum have shown how this may happen over the *longue durée*, while non-western societies have also sometimes adopted or incorporated western patterns of timing with a marked haste.[22] *Ancien régime* France also underwent a period of increased temporal policing and catechising of the Catholic calendar, drinking hours and market times; indeed the whole

21 J. Le Goff, *Time, work and culture in the Middle Ages*, Chicago 1980; G. T. Moran, 'Conceptions of time in early modern France: an approach to the history of collective mentalities', *Sixteenth Century Journal* xii (1981), 3–19; S. A. Epstein, 'Business cycles and the sense of time in medieval Genoa', *Business History Review* lxii (1988), 238–60; J. Scattergood, 'Writing the clock: the reconstruction of time in the late Middle Ages', *European Review* xi (2003), 453–74; M. M. Smith, *Mastered by the clock: time, slavery, and freedom in the American South*, Chapel Hill 1997. See also D. Gross, 'Temporality and the modern state', *Theory and Society* xiv (1985), 53–82.

22 P. Glennie and N. Thrift, 'Reworking E. P. Thompson's "time, work-discipline and industrial capitalism"', *Time & Society* v (1996), 275–99.

notion of policing may be seen as largely one of temporal control.[23] Newsprint may also have created a new sense of time, a new way of thinking that Benedict Anderson has deployed in his influential *Imagined communities*.[24] Such shifts in policies and technologies points to the fact that the calendar had been changed, and could be changed again. As the Italian historian Francesco Maiello has also shown, post-Tridentine France underwent a transformation in calendar use and in the patterns of religious observance.[25] Under Louis XIV and his successors, France experienced a process of temporal reorganisation, with fairs and markets coming under increased regulation and control, in their hours of operation as well as their dates. Clocks spread throughout towns, and bells continued to spread through the countryside. The growth of print culture also encouraged the use of what d'Alembert dismissively called the 'Calendrier Rustique', an assortment of almanacs by means of which countryfolk 'apprehended the time when they must sow, plant, cut the vine, etc', generally by 'false rules ... founded for the most part on the influence & the aspect of the moon and the planets'.[26] For many among the social and educated elite, a scientific revolution led to a fascination with watches, clocks and, for some, an obsession with self-improving daily routines for reasons of health, manners or education. Similarly, as the *Encyclopédie*'s entry for 'Calendrier' made clear, the calendar was not a fixed object, but a historical creation that differed by nation or religion. The hours of the day were similarly varied, whether mean or solar, or Italian. Almanacs, including Sylvain Maréchal's deliberately impious *Almanach des honnêtes gens*, as well as less provocative titles, suggested a certain playfulness with the way in which the calendar could be approached. In this context, a sustained effort by the republican authorities may have succeeded in altering apparently embedded customs.

Time, history and the Revolution

French intellectual thought at this time also placed a heavy emphasis on a chronological view of the world, and may even be said to have defined the historicist sensibility and history as a discipline distinct from antiquarianism.[27] In particular, historical writing of several kinds provided one of

23 Maiello, *Storia del calendario*; Beck, *Histoire du dimanche*; A. Williams, *The police of Paris, 1718–1789*, Baton Rouge 1979; J. Thomas, *Les Temps des foires: foires et marchés dans le Midi toulousain de la fin de l'ancien régime à 1914*, Toulouse 1993.

24 Anderson, *Imagined communities*.

25 Maiello, *Storia del calendario*.

26 'Calendrier rustique', *Encyclopédie*, ii. 555.

27 N. Hampson, *The enlightenment*, London 1968; Jones, *Great nation*, 186–92.

the central intellectual tools of eighteenth-century public and political culture. Following the great debate between the 'Ancients' and the 'Moderns', historical writing had been one of the engines that opened Louis XV's reign up 'to critical scrutiny' and, by mid-century, the discipline had retained and even sharpened its polemical edge. Again, the *Encyclopédie* reveals a profound interest in the workings and shaping of time, setting out the case for history as a triumphant march of Enlightenment in the 'Preliminary Discourse' (1751). Perhaps the most famous Enlightenment adventurer, Voltaire, consolidated his literary and philosophical reputation with his histories, notably *Le Siècle de Louis XIV* (1751), which also sought him favour with Louis XV, although his tenure as Royal Historian ended, as one recent historian notes, as a 'flop'.[28] He devoted the entry on history in the *Encyclopédie* to defending the discipline's contribution to knowledge; elsewhere, as in the *Essai sur les moeurs*, he charted the complex course of progress and barbarism.[29] Works such Montesquieu's *De L'Esprit des lois* (1748) used analysis of past regimes or societies as a way of critiquing the present. Historical and ethnographic comparisons were used to show how society passed through various stages on a progressive path towards Enlightenment.

The historicist sensibility can also be seen in works such as Buffon's *Natural history* (from 1749), which ranged chronologically as well as geographically, challenging the Church's account of the age of world. By such methods, time could be charted more accurately by fossils and geological features rather than the Bible.[30] The influential Scottish historian William Robertson, who was translated and published in France, found the production of histories almost overwhelming: 'the universal progress of science during the last two centuries, the art of printing, and other obvious causes, have filled Europe with such a multiplicity of histories, and with such vast collections of historical materials, that the term of human life is too short for the study or even the perusal of them'.[31]

Other nascent disciplines also placed a special importance on temporalities: Francois Quesnay stressed the importance of time to economic cycles, naturalists paid attention to the seasonality of flora and fauna, astronomers

28 Jones, *Great nation*, 188.

29 K. O'Brien, *Narratives of enlightenment: cosmopolitan history from Voltaire to Gibbon*, Cambridge 1997, 22–4, 43, and 'History and literature, 1660–1780', in J. Richetti (ed.), *Cambridge history of English literature, 1660–1780*, Cambridge 2005, 365.

30 A brief account of the influence of historicist writing can be found in Jones, *Great nation*, 186–8.

31 William Robertson, preface to '*The history of Charles V*', in F. Gilbert (ed.), *The progress of society in Europe: a historical outline from the subversion of the Roman Empire to the beginning of the sixteenth century*, Chicago–London 1972, 3.

made increasingly accurate timings of the motions of the planets, and geological time had begun to challenge the authority of biblical chronologies.[32] The leading journals and publications of the Enlightenment, impressed by clockwork, precision and historical enquiry, reveal a fascination with the science of chronology or chronometry. In 1753, for example, Barbeu Dubourg argued that chronology arose from the same needs as geography and was even more important than that discipline, while the *Encyclopédie* devoted plates to the art of horology and commissioned the leading horologist of the day as a writer.[33]

The idea of progress, of course, was based on the idea of temporal development.[34] Communication and connections between peoples led to progress, while isolation and social breakdown led to the loss of knowledge over time, such as that suffered by the west following the decline of Greece and Rome. For men such as Turgot or Condorcet, the lesson of history was one of optimism, despite the ebb and flow of progress over time. Technological innovation, such as language, writing, libraries and transport links, ensured that human society was on an inevitable, if gradual, path towards progress. Such a view promised the endless intellectual and moral advancement of the human race, an achievement only endangered by 'ignorance'.[35]

This utopian spirit would deeply influence many actors on the Revolutionary stage, at least in terms of enabling its participants to imagine a better world and to believe that history was on their side. History was littered with examples of the tyranny of kings and foreign powers, but also the unavoidable progress of reason, as demonstrated by the gains that Liberty had made, most dramatically in 1789.[36] The apogee of the conception of 'progress' might be found in the life of Condorcet, particularly his final few months on the run from the Jacobin authorities because of his involvement in the Girondin 'faction'. Hiding in Paris in 1794, Condorcet turned to his writing and philosophical studies, charting the course of human society in his *Esquisse d'un*

32 S. Meysonnier, *La Balance et l'horloge: la genèse de la penée libérale en France au XVIIIe siècle*, Paris 1989.

33 B. Dubourg, *Chronographie, ou description des tems; contenant toute la suite des souverains de l'univers, & des principaux événemens de chaque siècle, depuis la création du monde jusqu'à présent en trente-cinq planches gravées en taille-douce, & réunies en une machine d'un usage facile & commode*, Paris 1753, 5–6. See also D. Rosenberg and A. Grafton, *Cartographies of time: a history of the timeline*, New York 2010.

34 R. Nesbet, *History of the idea of progress*, New York 1980.

35 G. S. Painter, 'The idea of progress', *American Journal of Sociology* xxviii (1922), 262.

36 *Programmes des professeurs de l'école centrale du département de l'Isère pour le cours de l'an VII*, Grenoble Year VII, 29–30, AN, F^{17}1338. For an examination of a slightly later period see J. K. Burton, *Napoleon and Clio: historical writing, teaching, and thinking during the First Empire*, Durham 1979.

tableau historique des progress de l'esprit humain and arguing that the study of the past taught that man could 'transcend history'. Its sole purpose was to reveal how reason led to freedom.[37] Although Condorcet had fallen foul of the Revolution's political twists, his view of the progressive nature of human society was widely shared among his contemporaries, and had informed much of the Revolutionaries' programmes from 1789. During the eighteenth century, as John Burrow suggests, the view that history was a series of successive stages had become a 'characteristic genre', evinced in Voltaire's *Essay on customs* (1756) and Jean-Jacques Rousseau's *Discourse on the origins of human inequality* (1755) and in the much-translated works of the Scottish Enlightenment.[38] The notion that history carried a message of progress emboldened the plans for national regeneration and the restructuring of society after 1789. Speeches in the Assembly and Convention warned of the perils of 'ignorance', the counterpart to superstition, while even relatively straightforward guides to Paris and its brothels made ironic play of the benefits brought by philosophical 'progress'.[39]

The idea of intellectual, moral and political improvement, it seems, had become something of a well-worn literary trope, and by the 1790s the past was not to be celebrated, but used as a salutary lesson. The classical world proved an exception: schooled typically in the *collège*, the rhetoricians of the Revolution, who prepared their many and lengthy speeches in advance, drew upon their classical education to create a new symbolic language for the republic. As Walter Benjamin noted, 'for men such as Robespierre, ancient Rome was a past charged with the time of the now which he blasted out of the continuum of history. The French Revolution viewed itself as Rome incarnate. It evoked ancient Rome the way fashion evokes costumes of the past'.[40]

In these ways, the revolutionaries would reject the medieval and early modern past, while seeking inspiration in the ancient world. While time might once have been imagined as circular, constant and predictably repetitive, this temporal map was torn up by the Revolution, with the demand that a new chart be drafted. Genuine moments of watershed are rare in history,

37 M. J. A. N. Caritat, marquis de Condorcet, *Esquisse d'un tableau historique des progrès de l'esprit humain, suivie de réflexions sur l'esclavage des nègres, par Condorcet*, Paris 1822; K. M. Baker, *Condorcet: from natural philosophy to social mathematics*, Chicago–London 1975, 349.

38 J. Burrow, *A history of histories: epics, chronicles, romances & inquiries from Herodotus & Thucydides to the twentieth century*, London 2009, 341–2.

39 L. V. Thiery, *Almanach du voyageur à Paris*, Paris 1789, v.

40 W. Benjamin, 'On the concept of history', http://www.marxists.org /reference/archive/benjamin/1940/history.htm.

but contemporaries recognised 1789 as one such, even if the true extent of the transformation and the nature of the forces unleashed by the Tennis Court Oath, the Declaration of the Rights of Man and of the Citizen, and the abolition of feudalism on 4/5 August, were not properly understood by their participants (if, indeed, they are today).[41] Metaphorically, the shape of time had shifted from the cyclical to the linear, and the symbolic language of the nation had fallen in line with the *philosophes*' notion of the natural world and human society, moving in one, improving direction.[42] The past could be packaged up as something that reason had allowed France to escape, and the nation could be reoriented around the future. The language of regeneration, the interest in educational schemes, and the rhetorical weight often placed on children as the future of France would all derive from this viewpoint, and found practical expression in the outpouring of almanacs, chronologies and various 'cartographies of time'. While the chronologies that could often be found in the front matter of republican almanacs, for example, linked the past to 'superstition' and 'barbarism', the Enlightenment interest in time and the science of history ensured that temporality continued to have its place in public discourse. Perversely, perhaps, during this moment of interruption of linear time, there was national interest in calendars and in what Sanja Perovic has described as the 'historicization of meaning'.[43] It was into this space that the new calendar was placed.

41 T. Tackett, *Becoming a Revolutionary: the deputies of the French national assembly and the emergence of a Revolutionary culture*, 1789–1790, Princeton, NJ 1996. See also Smith, 'Between discourse and experience'.
42 Rosenberg and Grafton, *Cartographies of time*, 13.
43 Perovic, 'Untameable time', 192.

2

The French Republican Calendar, 1793–1806: A Narrative Account

On 14 July of the fourteenth year of the reign of King Louis XVI (1790), one year after the fall of the Bastille, some 350,000 men, women and children gathered on the Champ de Mars in Paris to celebrate the great *fête de la fédération*. The rain that fell during the ceremony surely did not detract from the great sense of occasion shared by the huge crowd as Talleyrand, the bishop of Autun, celebrated mass, and the king of the French, Louis XVI, swore his oath of loyalty to the constitution before an altar to the fatherland. The festival, at once both ancient and profoundly new in its symbolism, marked the culmination of a month of events celebrating the 'federation' of France and the first anniversary of the Revolution.[1] Although at root a military ceremony in which national guardsmen gathered from across France to swear allegiance to the king and the national assembly, the festival came to symbolise 'joyous accord' for the French nation, expressing hope for the future and social and political harmony. The physical space of the Champ de Mars had been reshaped by the people of Paris in the '*journée* of the wheelbarrows', in which the earth of the old military parade ground was organised into great banks for spectators that would last until the Second Empire. The materials of the old regime, it might also be said, were being reshaped into the site of the new order of things. Time had become politicised, its shaping linked ever more closely to the nation's history.

The exact meaning of this liberty could not remain uncontested. Certainly the harmony that was sought in the *fête de la fédération* soon proved to be elusive. While the new government attempted to resolve the problem of public finances, bring order to the countryside and confront a growing aristo-

1 W. Doyle, *The Oxford history of the French Revolution*, Oxford 1989, 129; T. Tackett, *When the king took flight*, Cambridge, MA 2003, 37. On the Revolutionary optimism and feeling of the first year of the Revolution see Tackett, *Becoming a Revolutionary*. Some estimates place numbers of witnesses to the Paris pageant as high as one million.

cratic opposition, a form of rhetoric that emphasised the dawn of a new era gained momentum and greater influence.[2] War, counter-revolution, popular agitation and the king's recalcitrance further radicalised the Revolution and underscored France's political and social rupture with the past. Less than a year after the *fête*, Louis XVI had, in the unsuccessful flight to Varennes, signalled the beginning of the end of the monarchy. In September 1792 France declared herself a republic. Cultural transformation went hand-in-hand with this political transformation, as a self-consciously republican mode of dress, language and even deportment, sought to create a new nation, defined in opposition to the former – *ancien* – regime.[3] Then, in the autumn of 1793, in one of the most radical reforms of the revolutionary era, the Convention declared it to be retrospectively Year I of the French republic, ending centuries of use of regnal years, dispatching the Gregorian Calendar and constructing a new, republican calendar suitable for a regenerated nation. As David Andress has observed, the men of 1793 had 'seized revolutionary control of the very nature of time'.[4] This chapter provides a narrative of this history, charting the course of the republican calendar from its conception in the thinking of the Enlightenment to its death during Napoleon's empire.

Time, of course, easily eludes the control of even the most diligent administrator, and the twelve years during which the republican calendar was enforced witnessed the Terror, Robespierre's fall, the Thermidorian reaction, the political see-saws of the Directory and Napoleon Bonaparte's ascent to power. It would be difficult to to accommodate these political twists and turns on the blank slate of the new calendar, which instead took on ever-stronger republican associations. Abolished by Napoleon in 1806, the calendar then entered the realm of collective memory, acting as political shorthand for Jacobin sympathies and offering to the public ideologically resonant titles, such as Victorien Sardou's daring republican 1891 play, *Thermidor*, or Zola's novel of the birth of proletarian awakening, *Germinal*. Revolutionaries in the nineteenth and twentieth centuries self-consciously followed in the calendar's statement of novel reform, whether in the revived republican journal *La Père Duchesne* during the Commune or the Bolshevik calendar reforms.[5]

2 Baker, *Inventing the Revolution*.

3 For a recent restatement of revolutionary expectations see L. Hunt, 'The world we have gained: the future of the French Revolution: presidential address to the American Historical Association, 2003', *AHR* cviii (2003), 1–19; Baker, *Political culture of the new regime*; C. Weber, 'Freedom's missed moment', *Yale French Studies* xi (2001), 9–31.

4 D. Andress, *The Terror: civil war in the French Revolution*, London 2005, 291.

5 S. Kroen, *Politics and theater: the crisis of legitimacy in restoration France, 1815–1830*,

Although arguably planned as a politically neutral measure, the symbolic framework of the republican calendar could not help but overwhelm it.

Perhaps even more important, the narrative of the history of the Revolution could only become apparent in retrospective, with its terminology, key events and contours neatly defined by the techniques of historical prose. The calendar, and its programme of festivals, struggled to accommodate these changes, becoming ever more anachronistic as the distance from 1789 grew longer, more controversial, and distinct from individuals' experiences. In contrast to the calendar's attempt to provide France with a new cosmology, in practice its implementation was the result of a thousand administrative actions or inactions, acceptance or resistance in a million personal acts, and a never-ending printed or handwritten skirmish in the dating of personal correspondence, newspapers, almanacs, contracts, passports and forms. As Sanja Perovic suggests, time proved to be 'untameable', ultimately out of reach of the ideological grasp of the Revolution. Even more disappointing to revolutionary heirs of the Enlightenment, who saw the calendar as the means to regenerate and educate a nation, reform often stumbled in the face of local opposition, misunderstanding or plain contrariness: a pragmatic response to political hyperbole and the revolutionary rhetoric and discourse traced by historians such as Franèois Furet and Mona Ozouf.[6] The history of the calendar offers a window into the history of the creation of a nation, as imperfect, bureaucratic, mundane and locally determined as that might be.[7]

History and the rhetoric of Revolution

An appeal to history had constituted an essential element of the pre-revolutionary debate, which considered the state – and origins – of the nation. Between July 1787 and September 1788, more than a thousand pamphlets were published, mostly anonymously and cheaply printed. A sizable and influential proportion of these made use of historical arguments, often based on the supposed antecedents of the French nation among either the Gaulish or the Frankish tribes, and indeed historical arguments had surfaced earlier in the century with the writings of Boulainvilliers and in the aftermath of the Maupeou coup.[8] Pamphleteers compared these ancestral

Berkeley 2000, ch. iii; Baczko, 'Le Calendrier républicain'; E. Weber, 'About Thermidor: the oblique uses of a scandal', *FHS* xvii (1991), 330–42.

6 For a more extended critique of Furet and Ozouf's still-dominant interpretation see Byrnes, *Catholic and French forever*.

7 Perovic, 'Untameable time'.

8 K. Margerison, 'History, representative institutions, and political rights in the French

tribes either to the current aristocracy or the Third Estate as required for political and rhetorical purpose. The majority of the publications were known as 'patriotic pamphlets' by contemporaries, written by local *avocats*, who supported parliamentary constitutionalism, tracing their arguments for political legitimacy to Gaul, and usually drawing the radical conclusion that the Estates General rather than the monarch was the nation's most legitimate agent. For this group, the authority of the Estates General lay in historical precedents, which had only recently been swept aside by despotism and monarchical interference. As Dale Van Kley summarises, 'these partial harbingers of revolutionary ideology are relentlessly historical, reading French history as a crescendo of usurpations of national constitutional rights by the forces of "ministerial" if not monarchical "despotism"'.[9] Other pamphlets, which found the origins of France within the Frankish kingdom, supported the 'ministerial' position and attacked the 'patriotic' defence of privilege.

Although the past was important in this debate, views of the future also played an increasingly important role. Pamphlets such as *Examen impartial de cette question*, published by a 'citizen of Besaneon' in 1789, gave an intimation that a new era, separate from the burden of the past, was about to begin. The author reflected that 'a philosophical spirit' has achieved much in the last century. The shadows of ignorance were beginning to fade and Enlightenment was spreading, leading him to have cause to hope that this 'mass of Enlightenment' would be augmented by ensuring the impossibility of a 'return to barbarity'. Throughout Europe, the arts of navigation, of commerce, of printing and the creation of the postal system had 'multiplied the relations and communications amongst men'. Within this context of the spread of rational reform, he hoped that the Estates General would assist the general retreat from unreason and provide the Third Estate with a voice. The author still drew on the past, finding authority for such an assembly as originating in Charlemagne and gave examples of comparable diets throughout Europe.[10] The authors of similar texts appealed to history and to the power of reason, suggesting that France might adopt a balanced constitution similar to that of England.

pre-Revolution, 1787–1789', *FHS* xv (1987), 68–98; H. A. Ellis, *Boulainvilliers and the French monarchy: aristocratic politics in early eighteenth-century France*, Ithaca 1988.

9 D. Van Kley, 'From the lessons of French history to truths for all times and all people: the historical origins of an anti-historical declaration', in D. Van Kley (ed.), *The French idea of freedom: the Old Regime and the Declaration of Rights of 1789*, Stanford 1994, 81.

10 *Examen impartial de cette question: comment et dans quelle forme doit-on procéder aux délibérations de l'assemblée prochaine des Etats-généraux?*, Besaneon 1789.

The employment of the historical method by pamphleteers could not answer all the demands of the brave new world created in the summer of 1789. Despite this careful construction of antiquarian links with the past by both reformers and the monarchy, the events of May–August of 1789 were unprecedented. The Estates General became, it believed, nothing less than a national assembly, with the self-declared task of providing France with a constitution. Spurred on by hunger and rumours of plots and aristocratic reprisals, the people of Paris stormed what had become the brooding symbol of despotism, the Bastille, looking for arms and gunpowder.[11] Likewise, the countryside began to erupt in waves of revolt. On the night of 4 August the Assembly renounced its privileges, overthrowing the perceived feudal structure of France, and consolidated the legal and rhetorical revolution with the Declaration of the Rights of Man and the Citizen. Then, on 5 October, the Parisian crowd marched on Versailles, overwhelmed the guards and brought the king and queen – the 'baker and baker's wife' of popular imagination – back to the Tuileries Palace in Paris. Few could avoid feeling as though they had entered a new era, whether for good or ill, one that had somehow escaped the precedents of the past.

Almost from the first meeting of the Estates General people had begun to talk or write of 'The First Year of Liberty', underlining the real sense of a new dawn ushered in by the drama of 1789.[12] Journals and private correspondence testify to the new style of dating, showing a popular sense of a new era, and the willingness or need to think about time. Mastheads of newspapers, such as the *Moniteur*, soon carried the title 'The First Year of Liberty', and the events of 1789 were quickly recognised as a 'Revolution': a fresh beginning in the history of the world, rather than a cyclical event. Revolutions, Keith Baker explains, were understood as 'sudden changes in the political order of a state, without any connotation of a return to an earlier point'.[13] It was a chance for the state to be regenerated. This opportunity for rebirth and for regeneration, as Ozouf has highlighted, was one of the central themes during the course of the Revolution.[14] A new era had begun and an old era, the *ancien régime*, was

11 The people also stormed the Paris Observatory.

12 The concept of the 'year of freedom' was also used by English revolutionaries in the seventeenth century: S. Kelsey, *Inventing a republic: the political culture of the English Commonwealth, 1649–1653*, Manchester 1997, 93–4, 101, and D. Norbrook, *Writing the English Republic: poetry, rhetoric and politics, 1627–1660*, Cambridge 1999, 5.

13 K. M. Baker, *Inventing the Revolution*, Cambridge 1990, 218; A. Rey, *'Révolution': histoire d'un mot*, Paris 1989, 183; R. Reichardt and H.-J. Lüsebrink, 'Révolution à la fin du 18e siècle: pour une relecture d'un concept-clé du siècle des lumières', *Mots* xvi (1988), 35–68.

14 M. Ozouf, 'Space and time in the festivals of the French Revolution', *Comparative Studies in Society and History* xvii (1975), 372–84.

retrospectively defined and rejected. Despite his political fall, Condorcet continued to view 1789 as brilliant moment of optimism: 'A happy event suddenly opened a vast arena to the hopes of the human race; a single instant put a century's distance between yesterday's man and today's.'[15] It was into this temporal space that the 'eternal present' was born. The simple historicism of the Enlightenment did not account entirely for the Revolution. Because events had been so dramatic and the political transformation had been so profound, the national rhetoric of historical time, the neat and magisterial flow from the past to the present, no longer served. History, in the sense of tradition and heritage, became suspect.

This sense of dislocation and rupture is impossible fully to understand. In part, this is because such views of time operate internally, not just in the language of symbols and political rhetoric, but in more personal, psychological spaces. Lynn Hunt and Thomas Crow have suggested, for example, that the experience of Revolution and, especially, the execution of the king, created a sense of orphanhood and loss. For example, the huge popularity of the painting of the death of the child-hero, Joseph Barass, in 1794, can be taken to stand as an archetype of the fatherless son, in some way linked psychologically to the newly-created citizenry's loss and rejection of their king.[16] The influence of the 'odd-man out' of the Enlightenment, Rousseau, must also be brought into play. For him, the flow of history was not necessarily one of progress, but instead offered a radical alternative. Civilisation, as experienced under the *ancien régime*, did not bring amelioration or moral improvement, but separated man from the honest purity of the natural state. The abstract time of the Republican calendar, with its close associations with the natural world, the seasons and the stars, together with a system for simple, fraternal festivals, may be seen as a way of creating a new epoch – a modern present, separated from the past.

Such a sense of a new era suffused contemporary culture and writings of all kinds. Even texts such as the rakish, satirical guide *Les Bordels de Paris* (1790) made reference to the new epoch. The frontispiece to this work is dated the 'Second Year of Liberty' and much play is made of the immoral opportunities offered by the 'reign of liberty'.[17] The Revolution also figured prominently in more respectable, conventional texts, such as *Il Étoit temps ou la semaine aux*

15 Quoted in idem, 'Regeneration', in Furet and Ozouf, *Critical dictionary*, 538.

16 L. A. Hunt, *The family romance of the French Revolution*, London 1992; T. E. Crow, *Emulation: making artists for revolutionary France*, New Haven–London 1994. See also L. M. O'Connell, review of *Emulation*, H-France, Oct. 2006, http://www.h-net.org/reviews/showpdf.php?id=646.

17 *Les Bordels de Paris, avec les noms, demeures et prix*, Paris [1790].

événemens ([1789]), which provided an account of the events of the week of 14 July 1789. 'We have', its anonymous author suggested, 'achieved, in eight days, what force and the law could not effect for centuries.' It proposed a new assembly to be built as a temple of liberty on the ruins of the Bastille.[18] Other accounts often employed the traditional historical tools of chronology, anecdote and explanation to present the Revolution as a litany of events. Key episodes, such as the opening of the Estates General, the Tennis Court Oath, the storming of the Bastille and the night of 4 August were quickly recognised as points *en route* to what would become the republic, and became a series of commemorative moments that memorialised the past, almost as a classically-styled alternative to the Catholic Seven Stations of the Cross. Although nowhere stated, there is a sense that just as theologians characterised the Incarnation as both historical and existing outside of time, so the events of the Revolution were susceptible to chronological reckoning, but also became symbols of a new era, rather than simply political events. This secular, utopian theology quickly filled what were traditionally religious texts and generated a substantial body of almanac literature. Rarer, fictional narratives also made use of a near-canonical sequence of revolutionary tableaux, plotting characters' lives against the sequence of revolutionary *journées*, from the taking of the Bastille onwards.[19]

Temporal rhetoric infiltrated political culture and, especially, newsprint, which attempted to grasp these new, erratic and accelerated temporal rhythms. The periodicity and pervasiveness of the press was an eighteenth-century phenomenon, and one that was intensified by the experience of the Revolution. Arthur Young noted on 9 June 1789 that 'the business going forward at present in the pamphlet shops of Paris is incredible. I went to the Palais Royal to see what new things were published and to procure a catalogue of all. Every hour produces something new. Thirteen came out today, sixteen yesterday, and ninety-two last week'.[20] The hectic pace of events reflected in the production of newsprint continued, and even accelerated. As the journalist Antoine Joseph Gorsas wrote in August 1789, 'events follow one another with a rapidity that takes one's breath away'. Indeed, his 1791 account of the massacre at the Champ-de-Mars is written in the form of an hour-by-hour dispatch, placing the reader within the temporal ebb-and-flow of unpredictable events.[21] Such breathtaking rapidity of events demanded a

18 Anon., *Il Étoit temps ou la semaine aux événemens*, [Paris] [1789], 1.

19 M. Cook, *Fictional France: social reality in the French novel, 1775–1800*, Oxford 1993.

20 Quoted in Forrest, *Paris*, 56.

21 J. Suagnieux, 'Le Temps, l'espace et la presse au siècle des lumières', *Cahiers d'histoire* xxiii (1978), 313–34; E. Wauters, *Une Presse de province pendant la Révolution Française:*

new apprehension of time that was immediate and ever-changing. Gorsas suggested a need for a metaphorical 'new clock' to measure all that had passed, one that could divide time into short, but accelerating, intervals. The growth of the press and the increase in the reporting of national events arguably also contributed to a growing sense of nationalism, even of an 'imagined community' moving through time.[22]

As well as a concern with the historical context, the smaller moments of duration, those of hours, minutes and seconds, took on a new importance in the Revolutionary era. The speed of events and commentary, as Gorsas's reporting suggests, was unprecedented. New political forms also required a new measure of time, as the pamphlet *La Meilleure Manier de délibérer et de voter dans une grande assemblée* (1788) reveals. Written following the convocation of the Estates General, it examined how the assembly might 'make its resolutions with speed' and warned that if each representative were given three minutes to speak on an issue it would take nine hundred minutes, or fifteen hours, to consider 'the simplest question in the world'. If the assembly consisted of six hundred members, then these men would be confronted by debate lasting sixty hours. The attention of any assembly would be distracted by a great number of disparate speeches, 'weakening it, tiring it, leading it astray'. The author proposed a number of solutions, always alert to balancing equality in debate with the limited time available. The greatest importance was placed on completing a debate on an issue within one day, otherwise the assembly faced the 'greatest moral and political danger', since, 'it is in the interval between meetings [that] intrigues and seductions of all kinds can lead to the most unwelcome results'. Debate should be public, and it should take place within prescribed times, in contrast to private, secret moments of the day where old alliances and established networks of friendship, kinship and family could undermine the purity of the assembly's work.[23] *Relation de tout ce qui s'est passé en Provence* ([1789]) offered a chronological account of the Revolution in the south of France, emphasising the precise timing of events, recording the exact hours of the opening of meetings and their lengths. The creation of such a record gave permanency and legitimacy to the meeting of

journaux et journalistes normands, 1785–1800, Paris 1993, 183, 271, 355; D. Andress, *Massacre at the Champ de Mars: popular dissent and political culture in the French Revolution*, Woodbridge 2000.

22 C. Labrosse, 'L'Apport de l'étude des journaux de 1789 a la réflexion sur l'événement', in R. C. Grenoble (ed.), *Aux Origines provinciales de la Révolution*, Grenoble 1990, 258–9. See also his 'Le Temps immédiat dans la presse parisienne de 1789', in *L'Espace et le temps reconstruits: la Révolution Francaise, une révolution des mentalités et des cultures?*, Aix-en-Provence 1990, 110–20.

23 *La Meilleure Manier de délibérer et de voter dans une grande assemblée*, n.p. 1788.

the provincial assembly.[24] Such texts reveal the Revolutionaries' interest in the exact nature of time and its public role, ensuring that temporality was placed under the control of the 'national will' rather than of other entities, such as the Church, the aristocracy or the monarchy.

The reform of weights and measures

The 'Ur-text' of the republic of letters, the *Encyclopédie*, reveals a fascination with time and measurement and many articles (and especially their footnotes) also show a concern for its reform. Weights and measures, Diderot and d'Alembert believed, needed to be made universal and based upon a unified principle. They also favoured the decimal system, and even wondered if it should be applied to time.[25] Increasingly aware of the broad range of time-systems used in different ages and other civilisations, any account of the calendar seemed to imply its potential reform, saturated as it was with religious assumptions and irrational inheritances from the past. Agronomists, moral philosophers and Physiocrats, concerned with what would become the social science of economics, also questioned the need for the number of feast days and joined moralists in lamenting the loss of labour and the increase in drunkenness and unholy disorder that saints' days tended to promote despite the best efforts of generations of priests and policemen.[26] The lack of uniformity of other weights and measures led to countless problems for trade and science, but reform seemed to be slow in coming. Despite the existence of proposals from Condamine, Turgot and Necker, as Prieur noted, 'prejudices' and habit easily averted any reform.[27]

In 1788 the calling of the Estates General seemed to offer the opportunity to cut through this Gordian knot of inertia and local privilege. Partly in response to the demands made in some of the *cahiers de doléance* for regional measures based on the standards of the larger market towns, but also in recognition of the profusion of different weights and measures that prevailed across the different towns and provinces of France, a group was commissioned to examine the problem under no less a figure than the scientist Antoine

24 *Relation de tout ce qui s'est passé en Provence*, n.p. 1789.

25 'Decimal', *Encyclopédie*, iv. 668–70.

26 Williams, *The police of Paris*; Beck, *Histoire du dimanche*.

27 C.-A. Prieur, *Mémoire sur la nécessité et les moyens de rendre uniformes, dans le royaume, toutes les mesures d'étendue et de pesanteur [Texte imprimé] … d'en régler tous les multiples et les subdivisions suivant l'ordre décuplé … par M. Prieur ci-devant Du Vernois*, Dijon 1790, 3–4; K. Alder, *The measure of all things: the seven-year odyssey and hidden error that transformed the world*, New York 2002, 35.

Lavoisier on 27 June 1789.[28] Authority over weights and measures were among the noble privileges renounced during the juridical revolution of August 1789, clearing the way for reform. As Ken Alder writes, 'from that point onwards, a flood of proposals poured in from the members of provincial learned societies, under-employed state engineers and enthusiastic citizens of all stripes'.[29] These suggestions meshed with the reformist zeal of the day, giving a further impetus to reform not just of money or the cadastre (the starting points for the overhaul of the measures of the *ancien régime*), but the 'era' itself.

Condorcet, Borda, Laplace and Legendre now joined Lavoisier to form a Commission of Weights and Measures and began work on the details. It looked at first as though the Parisian standards would be imposed across the nation, an idea initially proposed by the scientist Lalande.[30] But in March 1790, perhaps at Condorcet's behest, Talleyrand warned that the current proposals for the reform of weights and measures had risen neither 'to the importance of the matter nor the aspirations of enlightened and exacting men'. As he argued, in a land freed from the arbitrary rule of a monarch and the irrational accumulations of the past, new measures ought to be derived from nature itself; only then would they serve all mankind, impervious to the ravages of time and enabling peaceful commerce and communication between all peoples.[31] In contrast, the Parisian system of length was based on the *toise*. This six-foot iron measure could be inspected by interested parties at the base of the staircase of the Grand Châtelet. Over time it had become somewhat bent, suffered from fire and had had to be replaced on at least one occasion. It was not the certain measure on which to erect a new edifice of weights and measures for the new nation. The legislature promptly accepted Talleyrand's proposal, including his desire that the new measures should form an interconnected system, in which all units of measurement could be derived from one fundamental and natural measure of length. Such a system, its supporters argued, would be both elegant and practical. The legislature then determined that the system would be a decimal one (the neologism 'metre' had been coined by Auguste-Savinien Leblond in May 1790).[32]

At first, the proposed reforms also took on an international dimension,

28 Alder, *Measure of all things*, 89.

29 Ibid. 88; R. E. Zupko, *Revolution in measurement: western European weights and measures since the age of science*, Philadelphia 1990.

30 Alder, *Measure of all things*, 89.

31 Ibid. 90. Alder notes (n. 42) that Talleyrand's language here owes something of a debt to Condorcet.

32 'a name so expressive that I would almost say it was French', enthused Leblond: ibid. 92.

suggesting that the claims for the promotion of harmony amongst peoples were not mere utopian rhetoric. Talleyrand corresponded with the British parliamentarian, Sir John Riggs Miller, who had introduced a similar bill to reform British measures; support also came from Thomas Jefferson, who at George Washington's request had begun to reform the measures of the sister republic. Jefferson agreed to ensure that his scheme would be coordinated with that of the French. Condorcet noted optimistically that the 'three most enlightened and active nations', Britain, France and America, would soon be using the same measures, all calculated from the length of a pendulum that oscillated within one second. But where would the experiment take place? Gravity varies slightly according to latitude, so its location would affect the pendulum's motion. Unable to agree on a neutral site, each power proposed different places for the experiment: Sir John chose London, and Jefferson chose his Monticello home in Virginia. The French suggested Bordeaux, as close to sea level as possible. In the face of these differences, Talleyrand attempted to negotiate a compromise at the 45th degree of latitude (or the 45th parallel, which Jefferson then hoped would 'become a line of union with the rest of the world').[33] Despite these efforts, on 19 March 1791, the commission recommended that the metre be based on one ten-millionth of the distance between the equator and the North Pole, calculated from a survey of the meridian from Dunkirk to Barcelona. Conscious that the Revolution offered universal hope for mankind, the commission argued that only a measure derived from the globe itself, rather than a localised oscillation experiment, could provide a suitable gauge for all the citizens of the world. Borda, the chairman of the commission, also drew attention to the question of time. How could one fundamental measure depend upon another, which may be subject to change? The Academy was at the time considering converting the day of 24 hours of 60 minutes of 60 seconds into a decimal day of 10 hours of 100 minutes of 100 seconds.[34]

The republican calendar

On 2 January 1792 the Legislative Assembly decreed that all 'public, civil, judicial and diplomatic acts' must be inscribed '4th Year of Liberty' and that the 'era of Liberty' began on 1 January rather than, for example, 14 July.[35]

33 Ibid. 93–4. Anti-French feeling also put paid to British reforms: J. B. Morrell, 'Professors Robison and Playfair, and the *Theophobia Gallica*: natural philosophy, religion and politics in Edinburgh, 1789–1815', *Notes and Records of the Royal Society of London* xxvi (1971), 43–63.
34 Alder, *Measure of all things*, 95.
35 Tackett, *When the king took flight*, 69.

War and counter-revolution combined to radicalise the Revolution, and the break with the past was confirmed in September 1792 when France declared herself a republic. The new Convention eagerly approved a new seal for its records, replacing the king's head with a woman wearing a liberty cap: Marianne, the future symbol of the republic was born. The seal was soon adopted by all branches of the administration and the Convention declared that all acts would be dated from Year I of the French republic.[36] As January 1793 approached, the Convention began to consider the problem of when the 'republican year' began. Should subsequent years start on the arbitrary date of 1 January, on the more inspiring anniversary of the fall of the Bastille or on the date of the creation of the republic? On 20 December 1792 the Convention demanded that the committee of public instruction present a report on 'the advantages that France would gain by bringing into accord the republican era and the common era' as soon as possible. The report would go much further than this rather limited command.

By this point the Commission comprised Antoine Arbogast, Claude Joseph Ferry, Charles Gilbert Romme and Claude-Antoine Prieur de la Côte-d'Or.[37] These were not mere political placemen but shared a sincere interest in science and the possibility for reform. Arbogast was a mathematician of some note and the rector of the University of Strasbourg; he would be appointed professor of calculus at the École Central in 1794.[38] In 1790, as an officer of engineers with longstanding scientific interests, Prieur de la Côte-d'Or had presented the national assembly with a proposal for the reform of weights and measures. He argued that there were five systems of measurement – numeration (which included not only abstract numbers but all systems of relation, such as carats), time, *éntendue*, weight and money – and proposed reforms for the final three, noting that Europe already had common standards for numeration and time. None the less, he suggested that the division of time could be improved, or brought into line with other systems and be decimalised, making calculation and accounting easier for science, finance and commerce. The new political realities also gave impetus to change. Since the feudal system had been overthrown, the 'irons of despotism destroyed' and 'our great work of regeneration' had begun, it was time, Prieur argued, to change everyday measures as well. People would welcome this advantageous change 'with docility'. In this short paragraph in Prieur's

36 Hunt, *Politics, culture, and class*, 89–90; M. Agulhon, *Marianne au combat: l'imagerie et la symbolique républicaines de 1789 à 1880*, Paris 1979, 29.
37 Claude-Antoine Prieur de Vernois (1763–1832) was commonly known as Prieur de la Côte-d'Or after his native department: Prieur, *Mémoire*, 4, 7–8.
38 Louis Franeois Antoine Arbogast, 4 Oct. 1759–8/18 Apr. 1803.

Mémoire on weights and measures, he encapsulated the Revolutionaries' ambition to reform and regenerate society, an aim that was at once practical, political and moral.

Romme shared this utopian vision. A keen mathematician, sympathetic to Rousseau's fashionable educational theories, Romme had been tutor to the Russian count Stroganoff, but returned to Paris at the start of the Revolution, where he helped to found the revolutionary society, the *Amis des Lois*. After a short spell as a farmer in his native Riom, which allowed him to indulge his agricultural interests, he was elected to the Legislative Assembly and then to the Convention, where he joined the committee of public instruction.[39] The committee's minutes do not disclose who first proposed extensive calendar reform, rather than a simple 'harmonisation' of the republican and Common Eras, but as the minutes for 21 December 1792 reveal, the committee was discussing reform of weights and measures when time became entwined. It seems likely that Romme took the initiative to use the Assembly's decree as a starting point for a wholly new calendar; he undoubtedly profoundly influenced its shape and ambition.[40]

Romme did not work alone; he took care to consult the best and most sympathetic minds available. Jérôme Lefrancais de Lalande, the noted atheist, astronomer and mathematician, recalled that 'On the 31 January 1793, Romme wrote and asked me to confer with the members of the committee of public instruction, to reconcile the common era with the republican era. On the 8 March he wrote to me: I have received your paper, I will make use of it for the work that the Committee has charged me.'[41] Romme's report also notes the assistance given by fellow mathematicians, astronomers and deputies Pingré, Lagrange and Gaspard Monge, while Pierre Simon Laplace added his support for the decimalisation of time.[42] Undoubtedly, Maréchal's *Almanach des honnêtes gens* also provided the committee with a model for the new calendar's structure, one that was itself heavily indebted to the calendars of ancient Rome and Egypt.[43] Initially, Romme's other duties

39 Stroganoff returned to Russia in 1790 and Romme began to farm in Riom before joining the Legislative Assembly: *Les Stroganoff: une dynastie de mécènes*, Paris 2002, 37, 120–1.

40 Friguglietti, 'Social and religious consequences'. Romme's biographer, Alessandro Galante Garrone, agrees with this assessment of his centrality to the creation of the calendar, arguing that Romme soon went to work, 'scrupulous and obstinate' as ever, and, although he did consult several scientists, he was soon the only committee member occupied by the calendar reforms: *Gilbert Romme: histoire d'un révolutionnaire, 1750–1795*, Paris 1971; Friguglietti, 'Gilbert Romme', 13–22, and 'Social and religious consequences'.

41 Bibliothèque de l'Observatoire de Paris, MS B 57.

42 R. Hahn, *Pierre Simon Laplace, 1749–1827: a determined scientist*, Cambridge, MA 2005, 107.

43 Dommanget, 'Sylvain Maréchal', 301–34.

as a member of the National Convention intervened. In April 1793 he travelled to northern France as republican representative-on-mission with Prieur de la Côte-d'Or. It is possible that Prieur and Romme discussed the calendar reforms during this expedition to Calvados – the minutes of the committee of public instruction certainly record no discussion in his absence. Federalist rebels captured them in June, and imprisoned them in Caen.[44] They were released in late summer and returned to Paris, where Romme finalised the calendar during August and September.

Romme presented his report to the Convention on 20 September, with the hope of introducing the calendar on the following day's autumnal equinox, although it transpired that it could not be discussed until 5 October. The report concluded that the new year would begin on the observation of the autumnal equinox in Paris and would be numbered from the foundation of the republic in September 1792. The year would consist of twelve months, divided into three 'weeks' of ten days apiece; the remaining days would be grouped into a series of festivals or holidays at the end of the year. The *Moniteur* reported that each day would be dedicated to the republic, such as 'jour de la cocarde' or 'pique' and 'all that brings to mind the religious era would be abolished'.[45] Decimal reform of the hours of the day was also envisaged. Although the committee had been unable to decide on the names of the months, Romme included a table proposing names based on the Revolution (including 'The Mountain').[46] Finding these too historically and politically specific, the Convention passed the calendar into law without specifying new names for the months.[47] Two days later, the Convention

44 Georges Bouchard suggests that the calendar was discussed at this point: *Un Organisateur de la victoire: Prieur de la Côte-d'Or: membre du comité de salut public*, Paris 1946, 175. Garrone disagrees: *Romme*, 329. The evidence is at best circumstantial, although the two prisoners had the time and facilities to plan the project in the Château de Caen.

45 Romme had returned to Paris from Normandy on 5 Aug. He discussed calendar reform in the committee of public instruction on 19 Sept., where it was agreed that the hours of the day would be decimalised, but that this would not be compulsory until the 'Year III': *Procès-verbaux*, ii. 6, 438–40.

46 *Régénération, Réunion, Jeu de paume, la Bastille, Peuple, la Montagne, la République, l'Unité, la Fraternité, la Liberté, la Justice, Égalité*. The days of the *décade* were the classical and martial: *Niveau, Bonnet, Cocarde, Pique, Charrue, Compas, Faisceau, Canon*. The intercalary 'épagomènes' were *Adoption, d'Industrie, Récompenses, la Paternité, la Vieillesse* and, every ten years, *la Révolution*.

47 The rejection of specific names conformed to the Académie des Sciences opinion that measures should be derived from nature in order to avoid arbitrariness, exclusivity to one nation, or change from another 'revolution in the order of the world': *Rapport fait à l'académie*, 19 Mar. 1791, quoted in Académie des Sciences, *Sur l'Uniformité et le systême général des poids et measures: rapport et projet de décret présentées à la Convention nationale par*

Table 1

The seasons and months of the republican year

Autumn		
vendémiaire	month of vintage	September/October
brumaire	month of mist	October/November
frimaire	month of frost	November/December
Winter		
nivôse	month of snow	December/January
pluviôse	month of rain	January/February
ventôse	month of wind	February/March
Spring		
germinal	month of germination	March/April
floréal	month of flowering	April/May
prairial	month of meadows	May/June
Summer		
messidor	month of harvest	June/July
thermidor	month of heat	July/August
fructidor	month of fruit	August/September

declared that government officials would take every tenth day off, rather than every seventh.

The committee, now under the presidency of the former playwright and poet Fabre d'Églantine, was asked to return to the question of the names of the new months. Fabre d' Églantine presented the report on this subject on 24 October. The Convention approved of his work, and passed the new system of nomenclature, with its allusion to agriculture and the seasons, into law. The names of the months were grouped by the four seasons (*see* table 1).

Each day of the ten day week, or *décade*, took its name from a numerical sequence, running from *primidi* through to *décadi*. In keeping with the republicans's self-conscious debt to antiquity, the roots of the months were resolutely classical: *thermidor*, for example, was a compound of the Greek, *thermé* (heat) and *doron* (gift).[48] Each month consisted of thirty days, organised into three weeks, or *décades*, of ten days. The use of an intercalary days followed ancient practice, and the Athenian calendar, in theory if not in

le citoyen Arbogast et rapport fait a l'académie des sciences par les citoyens Borda, Lagrange et Monge, Paris 1793, 2.

48 Several traditional Eastern European calendars also followed this seasonal nomenclature. In Polish, for example, the eleventh month, *Listopad*, means 'falling leaves', while *Marzec*, the third month, denotes 'to freeze'.

practice, also used thirty-day months.[49] Similarly, the Egyptian calendar, as the *Encyclopédie* recognised, consisted of twelve months of thirty days and five days of holidays. Such symbolic links placed France as the heir to these illustrious forebears. The new months of the republican calendar also explicitly accorded with the zodiac and the understanding of the seasons, both of which were important considerations in the scientific world-view of the time.[50]

Fabre's report also settled other points. The year began on the observation of the autumnal equinox at Paris, when the sun moved into the sign of Libra, representing the harmony and balance desired by the Revolutionaries. By fortuitous coincidence this also marked the date of the declaration of the republic; a concurrence that may have clinched the decision to begin the calendar on the 22 September. The five remaining days of the solar year became *sans-culottides* and were earmarked for festivals. Hours and seconds were also decimalised, firmly linking the calendar to the wider republican reform of weights and measures, and creating a republican day of ten hours, rather than twenty-four, with each hour lasting 100 minutes, or 10,000 seconds. Tellingly, the report also concentrated on the new calendar's potential contribution to the eradication of 'superstition' and the dominance of religion in the public sphere. On 4 *frimaire* II (23 November 1793) these decrees were combined and published in a final form, officially ending several centuries' use of the Gregorian Calendar in France.

The Jacobin republic and the introduction of the calendar, 1793–4

Despite the calendar's origins in the universalism of 1789 and the scientific concerns of Enlightenment with the ordering and measurement of the natural world, it quickly became associated in the popular mind (and by many historians) with the political aims and policies of the Jacobin republic, and most notably became indelibly linked to the regime's hostility to the Catholic Church, which intensified as the course of events became ever more radical. The summer of 1793 had seen the vicious civil strife of the Federalist Revolt and war in the Vendée, together with continuing conflict with external powers, and on 5 September the Convention, spurred on by *sans-culotte* activity, declared 'Terror' to be the 'order of the day'. In October the execution of Marie Antoinette powerfully underscored the end of the *ancien régime*

49 More probably, the Athenian calendar consisted of months of either twenty-nine or thirty days in duration.

50 J.-J. Barthelemy, *Voyage de jeune Anacharsis*, Paris 1788, 225.

as well as the lengths to which the new regime would go. Such radicalism also began to characterise the Church as an enemy of the republic, and anticlerical feeling became increasingly organised. In September, for example, the Paris section Panthéon-franeais had called for Sunday 'schools of liberty' to counter the 'fanaticism' of religious preaching, and demands began to find greater support in the Convention and among popular politicians (although, notably, Robespierre and his supporters remained opposed to the nascent dechristianising movement). The calendar, which perfectly symbolised the republic's aims of effacing the past and constructing a new, utopian future, offered itself as a lightning rod for such feelings. In the historian Aulard's assessment, the calendar was the most antichristian innovation of the entire Revolution, potentially uniting a rather incoherent programme of popular dechristianisation with elite secularism and deism. Its very existence offered evidence for a broader shift in religious views that 'had been accomplished, or was preparing itself in men's minds'.[51]

The new calendar also sat alongside other Jacobin policies, plans and administrative reforms. A raft of policies, including the Maximum, plans for poor relief, education and the creation of national 'factories', was meant to weave the Revolutionary agenda into the fabric of the new nation.[52] As the barriers between private and public life were torn down, the calendar can be seen as another weapon in the Jacobins' cultural arsenal and at a time when the use of aristocratic modes of dress or speech or the use of 'Sunday best' might attract the attentions of vigilant surveillance committees, 'good citizens' might wear a decimal timepiece, self-consciously greet each other according to the new dates and times or carry republican almanacs or playing cards.[53]

But the reality of the calendar's implementation was more complicated than this relatively coherent picture of it as a crude cipher of political allegiance allows; it engaged with broader issues and concepts of commerce, order, festivity and agricultural practice, as well as the reality of the ways in which France was governed by the thousands of communes, new departments and by the local influence of Jacobin clubs. As an administrative reform, the calendar had to be pushed through the nascent republican bureaucracy in ways that differed according to geography or local political temperament, as well as simple chance and contingency (such as access to a printing press or

51 Aulard, *The French Revolution*, iii.158; A. Soboul, *The French Revolution, 1787–1799: from the storming of the Bastille to Napoleon*, New York 1974, 345. Soboul saw the calendar as a 'decisive landmark' (p. 582) in the separation of Church and State.

52 J.-P. Gross, *Fair shares for all: Jacobin egalitarianism in practice*, Cambridge 1997.

53 For an example of such cards see Alder, *Measure of all things*, 268.

Figure 1. Debucourt, 'Calendrier républicain', Paris 1794 (British Museum, 1929, 608.7).

stock of paper). In terms of the archival record, the calendar's initial impact appears to have been relatively uneventful, although its dissemination amongst the departments depended on the vagaries of the postal system (and on the actions of Vendéan or federalist rebels). Generally, officials adapted to the new system without much fuss, greeting the new calendar with just a few notes in the minutes of a department's central administration. Certainly, there were far more shocking laws to be implemented. Most departments adopted the new system relatively swiftly, although it sometimes took a few weeks for the full details of the calendar to arrive or be understood: in Toulouse, for example, republican dating was used from 1 November, but it took twenty days for Fabre d'Églantine's more poetic names of the month to replace the more prosaic '1er mois'.[54] In Besaneon, in the Doubs, the secre-

54 J. de Rey-Pailhade, 'Étude historique sur l'emploi du calendrier républicain et sur le temps décimal à Toulouse pendant la Révolution', *Bulletin de la Société de Géographie de Toulouse* iv (1908), 436; he may have been helped in this task by the poet and political journalist André Chénier.

tary of the *Justice de paix* began dating his minutes '4e an de la Liberté' from 7 June 1792. 'L'An premier de la République franeaise' made its first appearance on 29 September of that year. The republican calendar itself was used from 16 *brumaire* (the tribunal had not met since 29 September).[55] Closer to Paris, the new form of dating appeared in the registers of the canton of Beaune-en-Gatinais in the Île-de-France the next day, while in the Midi, Avignon began to use the new method of dating from the 20th of the second month (20 *frimaire*).[56]

Although clerks and other administrators could check the details of the new system against Paris publications such as the *Moniteur* or messages from central government, inaccurate printing and editing of departmental almanacs sometimes caused problems. Not only were the spellings of the months sometimes incorrect, but some could be in the wrong order. (*Ventôse* and *pluviôse* seemed to suffer unduly from this complaint.) Perplexity about the detail of the reforms was not uncommon, and such errors led to unease. Indeed, these represented one of the key early reactions to the reforms, and the Convention received complaints and notes of concerns from a number of departments. Attuned to military threats, the watch committee in Alsace, for example, wondered if there was some counter-revolutionary plot at hand: were the republic's enemies scheming by means of confused calendars to disrupt military affairs?[57] The military implications of the new system also led to complaints from the army, particularly about a lack of printed instructions. The time taken to calculate new astronomical tables led to a delay in the Bureau des longitude's publication of the *Connaissance des temps*, an important series of the longitudinal tables and tidal times that were vital for marine activities. Points of detail could also be dressed in ideological terms: while one commune in Calvados approved of the new calendar as 'most convenient for the whole people', it was unsure whether the *sans-culottides* – the new holidays at the end of the year – could be used for judicial business. The commune remarked that a free people, having escaped the burden of festivity imposed by the Church, should in no way afford to have less time available than 'enslaved peoples', particularly at a time of war.[58]

Government employees found their old weekly routines upturned. Quite

55 AM Besaneon, I 3 1.

56 R. Simon, 'La Révolution vécue par un bourg d'Ile-de-France: Beaune-en-Gatinais, 1789–1799 d'après les registres municipaux', *Revue d'histoire moderne et contemporaine* xviii (1971), 253. Beaune-en-Gatinais received the name Beaune-la-Rolande on 27 Aug. 1823: De Rey-Pailhade, 'Sur l'emploi du calendrier républicain', 440.

57 AN, F^{17} 1008 D.

58 AN, D III 37.

apart from the restrictions imposed on what could take place on the *décadi*, it is clear that the ten-day *décade*, rather than the hebdomadal week, began to structure the routine of a great many committees, juries and tribunals. The Convention decided on the '21eme du premier mois de la première année de la République' to receive petitions only on the tenth, twentieth and thirtieth of each subsequent month.[59] The *officier municipal* for the Paris Commune, warned that the assemblies for October, November, December and January would be arranged according to the *décade* and would not take into account the old days of the week.[60] For some, the hours of the day were also changed. Concerned at the lack of progress in defining other weights and measures, Lavoisier demanded that the Commission of Weights and Measures met on the second, fifth and eighth days of each *decade* at '7 *heures décimales* precisely (4 hours 48 minutes in the afternoon, old style)'.[61] Such novel arrangements were not limited to the Paris Commune and Convention. In *thermidor* of the Year II, the Nantaise 'tribunal de commerce' decided to meet according to the ten-day week, since their habit of meeting on Mondays, Wednesdays and Saturdays, perpetuated 'the former era'.[62] The department of the Côte-d'Or ordered all civil and political assemblies to adopt the new system. Fairs and markets would now take place in a pattern congruent with the new calendar, and all general councils, tribunals and judiciary meetings had to observe the *décadi*. Schools were to cease teaching on the fifth and tenth days of each *décade*.[63] A member of Dijon's surveillance committee said that he believed it to be:

> necessary to watch [those] who tend to work the day of the *décade* and rest on the day formerly known as Sunday. As a result, a motion is to be passed that all the members of the committee will themselves keep an eye on the citizens who seek to pour scorn on the Convention by paying no heed to the days intended for repose.[64]

59 *Le Moniteur*, 21 *vendémiaire* II (12 Oct. 1793).

60 *Journal du matin des Amis de Liberté et de l'Égalité*, DCCCVC (Year II), 4.

61 Lavoisier to the Commission of Weights and Measures, 22 of the first month, Year II (13 Oct. 1793), *Procès-verbaux*, ii. 387–8. This may be as much an indication of bureaucratic efficiency as evidence of Jacobin sentiment on Lavoisier's behalf.

62 Although the days were altered slightly in the Year VI, a pattern derived from the republican calendar predominated until the Gregorian Calendar was reinstated by Napoleon; Register, *Tribunal de Commerce*, Nantes, 1791–1849, entry for 24 *thermidor* II (11 Aug. 1794), AD Loire Atlantique, L 2113*.

63 *Arrêté du directoire du département de la Côte-d'Or concernant l'application de la division du nouveaux calendrier, aux époques des assemblées civiles et politiques*, Oct. 1793, AD Côte-d'Or, L 1660.

64 Register of the *Comité des surveillances*, 12 *floréal* II (1 May 1794), ibid. For similiar provisions in the Dordogne see H. Labroue, *L'Esprit public en Dordogne pendant la révolution*, Paris 1911, 94–102.

The proposal was passed unanimously. The fears revealed in this report were especially strong in border regions and those troubled by the signs of counter-revolution, where the calendar was quickly seized upon as an important means to erase such vestiges of the past and to measure the extent of republican regeneration. In the Somme, for example, the departmental authorities collected a series of notes from various administrative bodies and town councils on their implementation of the calendar, while in Amiens administrators were encouraged 'to follow the law promptly' so that the citizens could enjoy the benefits that would accrue from 'an augmentation in the number of days that public functionaries work'.[65] André Dumont, *en mission* with Franeois Chabot (who was also a constitutional cleric) to the Somme, seized on the dechristianising potential of the decree of 5 October and announced that 'festivals and Sundays no longer exist'. Priests who continued such 'hypocrisy' would be arrested and their religious paraphernalia burned.[66] In the first months of the calendar, its enforcement largely depended upon the enthusiasm of such *représentants en mission* and their views of the religious and political sympathies of the locals.[67]

The republic's finances also had to adhere to the new calendar. All authorities had to supply the central government with taxes and accounts, and these accounts needed to conform to the new timings. Bordeaux, for example, had provided the Trésorerie National with new accounts in October of each year, but were now required to provide accounts at the start of the republican year (a shift of a few weeks), adjusting payment accordingly. Similar adjustments took place across France. 'This new advance', the Trésorerie admitted, 'would present several difficulties in the first instance, but with a little thought you will understand perfectly.' The need to adopt the new calendar 'had to be incessantly addressed'.[68]

The military offered one important answer to this requirement. The rapidly-expanding army played an important role in the education of Frenchmen about their new rights and duties as citizens, and conscripts and volunteers received a great deal of propaganda and political education, including on the calendar. Since orders of battle and military reports followed the new calendar, familiarity with it was of paramount military importance.

65 AD Somme, L 421, 23 *brumaire* II (13 Nov. 1793).

66 AD Somme, L 421, '5 jour de la pres. mois, an 2'.

67 For example, Bô's efforts in the Loire-Inférieure: J. Bô, 'Le représentant du Peuple à Nantes, près de l'armée de l'ouest, & dans les Départements en dépendant, aux habitans des campagnes', 4 *messidor* II (22 June 1794), AD Loire Atlantique, L 99.

68 *Copie de la lettre écrite par les commissaires de la Trésorerie Nationale, au receveur du district*, '6eme jour, 2eme mois, an II' (27 Oct. 1793), AD Somme, L 421.

Regiments received primers on the new calendar, soldiers were paid according to its pattern and military routine conformed to it.[69] The same was true for the navy. An English prisoner-of-war, Major Watkin Tench, noted that the 'allowance of every person in the fleet, without distinction, [...] like everything else is *decimalized*, or regulated by periods of ten days' and commented on the attention given to *décadis*, which were 'distinguished by a more than ordinary chanting of republican songs, a display of the tri-coloured flag on the tops of churches in the town, and by a party of officers going on shore'.[70]

As the republic expanded its geographical influence, so did the new calendar – the new measure of time followed in the footsteps of the Revolutionary armies. The Batavian republic (as the Dutch republic was renamed by the French), marked the 'first year of liberty' by the publication of *La Breviaire du républicain, ou les loisirs d'un sans-culotte en Paris* in 1795, or 'An I. de la Liberté Batave', as the frontispiece read. The volume printed instructions for the 'French calendar for the use of the Bataves', explaining how the days were named after types of grains, trees and the fruits of plants and suggesting that these names would be found more 'useful' than the former commemoration of saints of the traditional calendar. Each month, the author clarified, had a different ordering principle (in the month of *nivôse*, for example, the days derived their names from the animal kingdom and from minerals) and a poem was provided to help with the memorisation of the names of the new months. The 'breviary' was illustrated by a common republican frontispiece, leaving the reader in no doubt of the calendar's secular intent: it portrayed a willowy *sans-culotte* trampling symbols of the Church. Republican armies also imposed the calendar on the Catholic heartland of the Italian peninsula, with copies published in the Cisalpine republic. A surviving example held by the British Library includes a full concordance to the old and new calendars, and a spirited effort to render the French months with their Greek origins into an Italianate form.[71] Here, republican military authorities were instructed to introduce *fêtes décadaires*, with predictable, and largely negative, reactions from the local population. In total, the calendar, together with the

69 Circular from the Minister of War, 12 *brumaire* II (2 Nov. 1793), AD Côte-d'Or, L 472; Bianchi, 'La Bataille du calendrier', 248. In 1799 conscripts were divided into classes according to their age at the start of the Republican year: T. Crowdy, *French Revolutionary infantry, 1789–1802*, Oxford 2004, 22.

70 W. Tench, *Letters written in France to a friend in London, between the month of November 1794 and the month of May 1795*, London 1796, 55, 63.

71 *Decadario per l'anno VI. della repubblica Francese, che comincia ... 21 Settembre 1797 col Colendario Romano*, Padua 1797.

new metric measurements, was imposed by officials on Switzerland, Belgium, parts of Germany, Italy and the Netherlands.[72]

Closer to home, the Convention paid some attention to the place of the calendar in private life. Although use of the new calendar was never made compulsory, it was wise to adopt it when one interacted with the official, or quasi-official, bodies of the state, such as the courts or Jacobin clubs. Newspapers and journals were required to use the new form of dating and, in a blow against the religious associations of the saints' days that populated such publications, almanacs were required by law to print the new calendar. Furthermore, Revolutionaries such as Romme believed that the good sense inherent in the reforms would recommend them to all citizens. Good citizens, of course, would also follow the proposals of the sovereign body, the National Convention. Where they came into contact with the government or judiciary, or took notice of official notices and proclamations, they would also inevitably discover the strange *nouveau style* (the abbreviation 'n.s' was commonly used) of measuring time. It was hoped that in time, as generations of French children, perhaps named Tridi Marat, Messidor Tell or Unité Indivisibilitée Décadi, were raised under the new system they would use the new calendar as first nature.[73]

Thermidor, 1794

Although the designers of the new calendar intended it to be a universal gift for all peoples and all time, it soon faced its first challenge of legitimacy. The overthrow of the Jacobin regime ten months after the introduction of the calendar presented a moment of potential historical re-evaluation. While the dismantling of the Terror's legislative machinery took some time to accomplish, Robespierre's fall from power in *thermidor*, Year II (July 1794) reshaped the political landscape and led to a re-emergence of the political right.[74] In Paris, gangs of disaffected rich young men – the 'Gilded Youth' – attempted to claim the streets of the city as their own with a mix of violence and ostentatious *ancien régime* dandyism, attacking the meeting halls of Jacobins and settling old scores. A right-wing press emerged, with powerful support from

72 H. A. Klein, *The science of measurement: a historical survey*, New York 1988, 125; T. C. W. Blanning, *The French Revolution in Germany: occupation and resistance in the Rhineland, 1792–1802*, Oxford 1983, 225.

73 P. Daumas, 'Prénoms et révolution, 1775–1825: propositions pour une nouvelle approche méthodologique', *Revue d'histoire moderne et contemporaine* xliv (1997), 109–32; Perovic, 'Untameable time', 1.

74 B. Baczko, *Ending the Terror: the French Revolution after Robespierre*, Cambridge 1994.

conventionnel such as Jean-Lambert Tallien, who accused the Convention of providing a roost for 'blood drinking tyrants'. In cafés and bars, rumours of the return of the monarchy began to circulate. Republicans were faced with the twin tasks of defending the Revolution from a swing to the monarchical right and of coming to terms with the fact that the Revolution had veered horrifically 'off course', from a vision of utopia towards a bloody dictatorship that imprisoned or guillotined thousands of its citizens. Together with ending the Terror and creating a workable constitution, the Thermidorians had to re-plot what might be termed the narrative of the Revolution: in Franєois Furet's words, 'they could not begin the Revolution once again, starting from 9 Thermidor by destroying the pages they had just written, as they had done in September 1792. They had to shoulder that history, and allocate the parts to be remembered and those to be forgotten'.[75] The year 1789 could no longer be seen as ushering in an uncomplicated era of hope, a consensual, timeless 'eternal present' or an escape from the unenlightened past. The future was now contingent and the recent past corrupted.

This reappraisal of the meaning of the republic and the Revolution had cast a newly critical light on Jacobin institutions, such as the calendar, and the men who introduced them. Pamphlets appeared condemning the calendar as an instrument of *terroristes* and *tyrans*. In one such tract, Lanjuinais, whose attacks on the *sans-culottes* forced him into hiding in June 1793, claimed that the majority of French citizens scorned the new calendar. The seasonal names of the months, he argued, were a 'perpetual lie' except for those who lived further south than Lyon, the calendar posed a great threat to commerce and since, 'neither man nor beast' could bear ten days of consecutive work, it could not be said to be in harmony with nature. The ties between the calendar and the Terror, particularly its links to dechristianisation, ensured that it risked being viewed as an objectionable instrument of a corrupt regime. After all, its end-of-year festivals were named in honour of the *sans-culottes*, the shock-troops of the Terror. Opponents of the calendar raised their voices in the Convention. In *pluviôse*, Faure, deputy for the Loire-Inférieure, demanded that the government should reconsider the 'alleged republican calendar' and postpone any decision on the nature of *fêtes décadaires*.[76] Elsewhere, official use of the calendar declined. Almanacs based on the Gregorian Calendar reappeared, and many communes no longer observed the *décadi*. Meetings of official bodies, such as the courts of many *juges des paix*, reverted to assembling according to patterns of the old

75 Furet, *Revolutionary France*, 152.

76 J. D. Lanjuinais, *Sur l'introduction du calendrier des tyrans*, Paris Year III; P. J. D. G. Faure, *Sur les Fêtes décadaires*, Paris Year III, 7.

seven-day week.[77] In a pragmatic rejection of one of the more extreme innovations of the Revolution, decimal time was abolished on 10 *ventôse* III (28 February 1795). It is doubtful whether it was still fashionable to wear a decimal timepiece in the months following *thermidor*, Year IV.

Yet, the bloody conclusion of the committee of public safety did not spell the end of the calendar. Publication of texts based on the republican calendar continued, even while the Gregorian made a return. Some, such as the *Almanach des prisons* (Paris, Year III) sought to make restitution for the errors of the *terroriste* regime. This text, prefaced by the republican calendar and a list of *fêtes décadaires* and their meanings, sought to satisfy the 'avid curiosity of the public' and to 'gather different anecdotes that will belong to history, and to steer the opinion of posterity on the persons who have strikingly adorned the theatre of Revolution'. The republican calendar was used within the anecdotes to date events, and one poem by an imprisoned schoolmaster even reflected that 'never will the sun of *brumaire*/ Have such charm for my eyes' as when he last saw his wife and child. The almanac was published in three volumes, each faced with woodcuts that reveal something of the attempts to understand the history of the Revolution. The first illustration shows heaps of decapitated heads placed in front of a guillotine, all under the banner, 'The Reign of Robespierre'. Although the clergy, the aristocracy and the parliament all have substantial mounds, the largest group of heads belongs to the 'people'.The second illustration depicts a meeting of a *sans-culotte* with a friend in prison. The caption reads 'I didn't expect to see you again. What are you doing in here?' The final print, an attempt to find redemption, portrays time as Chronos carrying Justice and is captioned 'time at last brings justice'. The sequence manages to portray the suffering of the people in the first figure and the human consequences of the Terror in the second, while the final illustration attempts to show resolution and recompense, without rejecting the republic. It is telling that the final illustration is the most abstract in design, with neither the horror of the guillotine and stacks of heads, nor the confusion, and even normality, of the second picture. In its abstraction, the calendar could in some ways escape the weight of meaning of the problematised recent past.

Classical allusion offered a way of escaping the difficulty of political or historical specificity and became increasingly important to the ways in which the calendar could be represented, not least in the attempt to organise a

77 P. Frand, 'Le Calendrier républicain dans les annuaires de l'an IV a l'an XIII', *AHRF* cciic (1994), 671–86; Shaw, 'Time and the French Revolution', appendix A; Meinzer, *Der französische Revolutionskalender*, appendices.

system of festivals that distracted attention away from the horror, historicity and contingency of the Revolution.[78] A *fête du calendrier* took place in *vendémiaire*, Year II (October 1794), at Arras, for example, which celebrated the abstract ideals of the Revolution, rather than its bloody or muddled content. As an aesthetic strategy, it offered a way of continuing with the calendar: while its abandonment may have made pragmatic, practical sense, such a *volte-face* remained ideologically awkward. To renounce the calendar, as Mona Ozouf argues, 'was to acknowledge that the events of the Revolution could be undone: this, a regicide Assembly could not tolerate'.[79] The Thermidorians were trapped, as Furet suggested, in a 'contradictory legacy', unable to reject a past that included liberty as well as Terror.[80] In the event, rather than undoing the gains of the Revolution, the Thermidorian Convention focused on preserving the republic, prosecuting a European war and bringing order to the Vendée. Work on a system of festivals and new weights and measures continued, albeit slowly. Yet, if Ozouf and Furet present a view of revolutionary government that posits a totalised republican ideology, in practical terms it lacked the time or ability to do much with the calendar. The situation was similar across France, as local administrators found the new calendar to be relatively low on their list of priorities. Where it did become an issue it did so not just because of its ideological weight, but also because of long-standing patterns of tension between centre and periphery, a matter of resistance to external interference from Paris.

The Directory, 1795–9

Although the calendar was promoted as a scientific advance, astronomers were aware of a serious scientific flaw: the new year began on the autumnal equinox observed at Paris, but to account for leap years, the decree of 1793 also stated that at the end of each four-year cycle 'one day is usually necessary, [it] is called the *Franciade*... The fourth year of the *Franciade* is called *Sextile*'.[81] However, the autumnal equinox did not exactly match this four-year cycle. On 29 *germinal* III (18 April 1795), Romme had gathered a group of mathematicians and astronomers to discuss intercalary reform to resolve this leap year question, and a few weeks later he proposed modifications that would correct this error. However, Romme fatally implicated himself during the Prairial uprising by calling for the release of imprisoned

78 *Almanach des prisons*, Paris Year III.
79 Ozouf, 'Revolutionary Calendar'.
80 Furet, *Revolutionary France*, 153.
81 Decree, 5 Oct. 1793, National Convention.

patriots and demanding improved food supply to Paris and permanent meetings of the sections. He was arrested in Normandy and brought back to Paris for trial. Found guilty, he committed suicide on route to the guillotine on 29 *prairial* (17 June 1796). His intercalary reforms were not implemented (and as a measure of anti-leftwing feeling, on 7 *fructidor* [24 August], the *jours sans-culottes* became the *jours complémentaires*).[82]

None the less, the calendar lived on as one of the remaining symbols of the republic. The new constitution even included the provision that the 'French era begins on 22 September 1792, the day of the foundation of the republic'.[83] Following the Fructidorian coup of Year VI, the Directory sought to reinvent and revive the calendar as part of a wider programme to create a republican citizenry and root out opposition. Belief in the need for reinvigorated revolutionary festivals and a renewed attack on Catholicism encouraged a series of measures defending the republican calendar and rooting out use of the Gregorian Calendar. On 14 *germinal* VI (3 April 1798), a decree on the use of the republican calendar was passed, followed that summer by a decree on the observation of the *décadi* and, in autumn, a further law on its observance. Throughout France, government officials, increasingly under the supervision of a more effective central authority, attempted to stamp out the old patterns of the calendar and public religion. Police patrols, reports and the energetic promotion of somewhat staid *fêtes décadaires* failed to create an organic, popular pattern of calendar observance.[84] To an even greater extent than it did during the Year II, the calendar became a central plank of the Directory's social and religious policy.[85] Meanwhile, the cult of 'theophilanthropy', a rather self-conscious, elite movement, very publicly adopted the calendar and its decadal festivals. Despite such support, as the historian Albert Mathiez made clear, the cult failed to become anything more than an intellectual, artificial innovation.[86] In contrast, massive departmental reorganisation of fairs and markets, the Directory's direct intervention in religious observance and the continued use of the calendar by newspapers made far greater inroads for the calendar into the popular imagination.[87]

82 M. Froeschle, 'A Propos du calendrier républicain: Romme et l'astronomie', *AHRF* ccciv (1996), 316.
83 Article 373 of the French constitution of the Year IV.
84 Shaw, 'Reactions to the French republican calendar'.
85 Livesey, *Making democracy*.
86 A. Mathiez, *La Théophilanthropie et le culte décadaire, 1796–1801: essai sur l'histoire religieuse de la Révolution*, Paris 1903.
87 Thomas, *Les Temps des foires*.

Napoleon and the reintroduction of the Gregorian Calendar

By the Year VIII (1799), when Bonaparte first came to share power following the coup of 18 Brumaire, it was conceivable that the calendar might become permanent. Joseph Bonaparte passed a number of measures to reinforce the *décadi*, the scientists Delambre and Méchain had determined the size of the globe for the first time, and French officials were busily imposing the new, and to many rather puzzling, measures of metres and kilogrammes. The growing French bureaucracy had become increasingly adept at using the calendar, and the new system even began to be adopted in some private correspondence.[88] Furthermore, as the eighteenth century drew to a close, calendar reform remained part of a public discourse, and a number of proposals suggested reforms to either extend the republican calendar, or align the dating of years with that of the new century.[89]

However, the calendar was too contaminated by politics, despite its claim to be a neutral, universal measure. A united Europe, under French dominion, was not yet in prospect, and Bonaparte was not interested in an unnecessary battle with the forces of French Catholicism, however personally disinterested in religion he may have been. On 18 *germinal* X (8 April 1802), Sunday replaced the *décadi* as the day of rest for government officials, a *de facto* recognition of the Gregorian Calendar's continued dominance of the quotidian. A number of objections to the republican calendar were also made, many of them practical. Most especially, it was a barrier to commerce. Savants could also point to the continuing problem of intercalation. Now installed as emperor, in the Year XIV (1805) Napoleon asked the scientist Laplace to prepare a report on the calendar. He proposed the reinstatement of the Gregorian Calendar and the abolition of metric weights and measures.[90] The Sénat accepted these recommendations, and the republican calendar (now titled the French Calendar) was abolished from 1 January 1806 in favour of the Gregorian Calendar. Other weights and measures remained untouched.

88 Stendhal [[M.-H. Beyle], *Oeuvres complètes de Stendhal*, ii, Paris 1923, 191; I. Tierder, 'La Calendrier républicain et ses incidences littéraires', *AHRF* ccxcii (1993), 259–67; J.- L. Ménétra, *Journal de ma vie: Jacques-Louis Ménétra, compagnon vitrier au 18e siècle*, ed. D. Roche, Paris 1982, 282–3; Maiello, *Storia del calendario*.

89 Some cognoscenti delighted to point out that the eighteenth century began in 1801, since there was no 'year zero'; others argued that this was too nice a distinction.

90 Laplace proposed retaining weights and measures as 'Napoleonic weights and measures', but this was rejected.

The move did not come as much surprise to anyone. *The Times* of London also concluded that the change in the calendar was a religious issue. Practical concerns were evinced. Earlier, in a report on events in France, the English paper noted that it had been informed on 30 July

> that the present French calendar will soon be abolished, it being found productive of endless inconvenience in mercantile transactions, in comparing dates of letters and bills of exchange, and possessing not one advantage in return, as it was not even astronomically just, and separated us from all the rest of Europe.

After the publication of the Sénatus-consulte, another story in the paper on 'The French Kalendar' took the change as an example 'of the unpopularity of those principles which gave birth and support to the system'. It recognised that a calendar could never be politically neutral and that the 'revolutionary kalendar was adopted, as one great means for the overthrow of Christendom and is most dear to its civil, political, and religious interests'. It argued that the republican calendar attempted a severe breach with the Christian order of things, by trying to sever the links between the seventh day and its divine ordination as a day of rest and worship.[91]

France largely shared this view of the return of the old calendar as the victory of Catholicism over godless republicanism. To celebrate the return of the Gregorian Calendar, the Théâtre Montansier staged a play entitled *Janvier et Nivôse*, a confrontation between the Revolution and Catholic France. The adventures of two young lovers, ciphers for the two competing calendars, provided the plot for the evening. Nivôse, inevitably weak and asthmatic, contrasted with 'young and gay' Janvier. For several years Janvier had been exiled, but returned with his friend Carême. He is welcomed with gifts of flowers and almanacs, and wins a final fight with Nivôse. The *Courrier des spectacles* praised the play for its celebration of the abolition of one of the Revolution's 'imprudent innovations', suggesting that 'the public has quickly seized on the timeliness of the subject and applauded all that relates to the superiority of the ancient calendar over the new'.[92] The allegory was clear.The calendar, relegated to bureaucratic habit or ostentatious, politically-limited Jacobinism, had withered on the vine of Napoleonic France. M. Décadi, his scientific rigour and utopian optimism all but forgotten, was vanquished by M. Dimanche, and the republican calendar entered the realm of collective memory, fiction and revolutionary myth, only to be resurrected with a certain self-consciousness by the Paris Commune of 1871.

91 *The Times*, 30 Sept. 1805.

92 *Courrier des spectacles*, 1 Jan. 1806, 2–3.

As a site of memory, the calendar had become indelibly linked to the sentiments of the Year II, even if it was most rigorously enforced during the Directorial period. While the subsequent history of the calendar is just as rich as its origins and the initial years of its use, its symbolic framework and its connections to the revolutionary moment have ensured that, for many, the calendar summons up visions of the guillotine or the *journées*, rather than years of bureaucratic use, or gradual and responsive local enforcement. While the later years of the calendar raise important questions, notably those related to work, leisure and state intervention in everyday life, we now return to its birth and the utopian visions of 1793, a period during which the calendar may be said to be bound up with the creation of a culture of republicanism.

3

Cultivating the Calendar: The Calendar and Republican Culture in the Year II

Time has not been kind to the papers of many of the French Revolutionaries; perhaps no others have been as widely dispersed as the manuscripts originally belonging to Charles Gilbert Romme, the creator of the French republican calendar. Although some of his papers, scattered through national and regional archives and libraries, can be consulted in France, many remain in Moscow. Following a sale by a Russian *émigré* in 1939, the Museo del Risorgimento, in Milan, purchased a small collection of Romme material and in consequence, among the papers of Camillo Cavour, Massimo D'Azeglio and Giuseppe Mazzini, three cartons containing seventy dossiers belonging to Romme can be consulted.[1] This collection reveals the range of Romme's intellectual interests, including an extensive pre-revolutionary correspondence with savants, notably botanists and astronomers, often with political commentary.[2] The collection also includes six notebooks in which Romme worked out the days of the new, republican months from *vendémiaire*, associating each one with rural produce or agricultural tools in a draft for the *Annuaire du cultivateur*, an almanac based on the new calendar that aimed to share the latest agronomic innovations with rural citizens, and in which hoes, apples and hares replaced saints' days and dominical letters.[3] The *Annuaire* embodied what was to become the agricultural and pastoral symbolism inherent in the republican calendar, a close connection between the year and rural life which attempted to make the calendar both simple to use and tied

1 A. Tchoudinov, 'Les Papiers de Gilbert Romme aux archives russes', *AHRF* cccvi (1996), 257–65; J. Godechot, 'Les Papiers du Romme', *AHRF* xxvi (1954), 63–73.

2 On Turgot, Oct. 1776: 'Un bon patriote doit s'en remettre au ministre sage, zélé et surtout désintérressé à qui notre roi donne sa confiance', quoted in Godechot, 'Les Papiers du Romme', 64.

3 C. Romme, *Annuaire du cultivateur, pour la 3e année de la république*, Paris [1794].

intimately to the rhythms of everyday life. Each month spoke of the wealth that derived from the land: the New Year began with the month of vintage (*vendémiaire*), and after the chill of the winter months, the months of germination (*germinal*), flowering (*floréal*), meadows (*prairial*) and harvest (*messidor*) promised agricultural abundance.

The calendar's rural associations were not merely a means of promoting the calendar in terms readily understood by the average citizen, but also had an important political dimension. Many strands of political philosophy argued that virtue could be encouraged by a harmonious relationship with the land and agricultural labour, and the Revolutionaries of the Year II sought to cultivate citizens' virtue and civic morality, just as the farmer tended his crops.[4] A route to civic regeneration could be found within the imagery of the natural world. This chapter charts in more detail the ways in which the republican calendar found cultural expression in the months following its introduction in the autumn of 1793. It shows how intimately many of these manifestations were linked to a particularly republican conception of the natural world, its relationship with moral regeneration, and also to nascent capitalism. Despite its designers' intention forever to link the calendar to the virtues of the natural world, such elite and highly literate allusions created a sterile cultural artefact, which failed to connect with the wider population. Where the calendar did create a spontaneous response was on the level of the object: the printed almanac, or decimal concordance, items of consumer capitalism which allowed the citizen to negotiate and relate their own daily life to that of the wider world and the nation.

The names of the months

Nomenclature had provided the first legislative crisis for the new calendar. Following a series of discussions in the committees and debating chamber of the National Convention, historical or overtly political schema were rejected and the calendar took its new names from the seasons and the months of the agricultural year. Romme had spent the spring and summer of 1793 (while not imprisoned in Caen) consulting some of the leading scientists of the day, eventually producing a calendrical system that linked the measurement of

4 P. Higonnet, *Goodness beyond virtue: Jacobins during the French Revolution*, Cambridge 1998, 114–15. On the importance of work and rest see V. Mainz, *L'Image du travail et la Révolution Franeaise*, Vizille 1999, and A. Mathiez, *La Vie chère et le mouvement social sous la terreur*, [Paris] 1927. On the pastoral form in literature see M. C. Cook, 'Politics in the fiction of the French Revolution, 1789–1794', *Studies on Voltaire and the Eighteenth Century* cci (1982), 233–340.

time to that of space, combining the committee of public instruction's task to consolidate the muddle of *ancien régime* measures and to harmonise the common era with that that of the republic into a larger project, more fitted to the new, rational world of the republic.[5] Romme described the republic's need for the new calendar, outlined its moral and scientific advantages, and showed how the existing names of the months, such as those that recalled the gods or 'tyrants' of the past (*janvier* – Janus, or *août* – Caesar Augustus) as well as the 'numerical' months from September to December, followed the Roman pattern of the year. The old calendar stood as a 'monument to servitude and ignorance'. He proposed a new set of names of the months, one 'neither celestial, nor mysterious', but derived instead from 'our revolution'. In this system, the new, republican year began with the month of 'régéneration' and ended with 'égalité'.[6] It made little or no reference to agriculture or the seasons.

For their part, the *conventionnel* seized upon the names of the months and began to argue about what they should be called; for example, the use of 'Mountain' for the month spanning February to March, while no doubt flattering to the current dominant political faction, also suggested a narrow political definition of the calendar which some worried might prove to be overly limited. The proposal presented to the Convention on 5 October failed to decide on the names of the months. It was still felt that the new republic demanded more expansive and suggestive labels than 'first month' or 'second day', despite the rational and logical purity of an ordinal vocabulary, and the Convention rejected such revolutionary terms as 'level', 'cap' and 'republic'.[7]

The debate over the names of the new months reveals the Convention's awareness of the political implications of the project. The calendar itself – its organisation, assumptions and language – was coded with these republican ideals and ambitions. It aimed to replace the errors and credulity of the past with a system that recognised the progress of the 'human spirit'.[8] Obscure saints days were to be replaced by a system informed by the seasons and the science of astronomy; national festivals replaced sectarian celebrations. The

5 C. C. Gillispie, *Pierre-Simon Laplace, 1749–1827: a life in exact science*, Princeton, NJ 1997, 154; *Procès-verbaux*, ii; Alder, *Measure of all things*; P. R. Hanson, *The Jacobin republic under fire: the federalist revolt in the French Revolution*, University Park, PA 2003, 68–70.

6 Beginning with the equivalent for September–October, the proposed months were régénération, réunion, jeu de paume, La Bastille, la Montagne, la République, L'unité, la fraternité, la liberté, la justice and égalité: C. Romme, *Rapport sur l'ère de la République*, Paris 1793.

7 Ozouf, 'Revolutionary Calendar'.

8 Romme, *Rapport*.

calendar also envisioned a systematic cultural programme, using the names of each day to educate the French about new agricultural methods, and supplying a festive framework through which fraternity and nationhood could be celebrated and reinforced. The calendar also provided the basis for a range of cultural responses, including poetry, theatre, ballet and the visual arts, mostly with a similarly didactic purpose. Although the calendar made claim to scientific purity and astronomical simplicity, it was not politically appropriate to offer a completely neutral measure of time (if such a thing were possible); it had to signify the new era that France had entered, providing a conceptual and moral framework for its citizens. As Romme claimed, 'the Revolution has given new vigour to the souls of the French; each day, every day it moulds them by republican virtues'; the calendar would also impress on them republican values.[9] Since the king's head had been removed and the Civil Constitution of the Clergy invoked, rearranging the calendar presented few problems to the Revolutionaries: the Convention's concerns with the reports of Romme and of Fabre d'Eglantine related to the calendar's meaning, rather than to any concerns about the practicality of the proposed reform.[10]

In contrast to the later image of the calendar as a divisive measure, in which the Catholic 'M. Dimanche' was set against a revolutionary 'M. Décadi', the proposals for the reform of the calendar were accepted by the Convention with relative ease because they lay at the heart of the republican vision of reforming not just the political, but the social and moral life of France's citizens.[11] The security of the republic would be based on such a transformation, and the arts were to be shackled to this republican purpose. Old, discredited images and assumptions had to be replaced by new, rational visions, and this shaped the symbolic framework of the calendar.[12] Fabre d'Eglantine's report explicitly stated that such propaganda was the aim of the reforms: the Gregorian Calendar, he suggested, 'filled the memory of the

9 'La révolution a retrempé les âmes des Franeais ; elle les forme chaque jour aux vertus républicaines': ibid.

10 P. F. Fabre d'Églantine, *Rapport fait à la Convention nationale dans la séance du 3 du second mois de la seconde année de la République franeaise, au nom de la commission chargée de la confection du calendrier, par Ph. France. Na. Fabre d'Églantine*, Paris Year II, 24 Oct. 1793; *Procès-verbaux*, ii. 693–713. The tone of the reforms relates closely to Sieyès's educational projects: R. Whatmore, *Republicanism and the French Revolution: an intellectual history of Jean-Baptiste Say's political economy*, Oxford–New York 2000, 96–8.

11 J. A Leith, *The idea of art as propaganda in France, 1750–1799: a study in the history of ideas*, [Toronto] 1965, and *Space and revolution: projects for monuments, squares, and public buildings in France, 1789–1799*, Montreal–London 1991; K. Emmet, *A cultural history of the French Revolution*, New Haven, CN 1989; Ozouf, *Festivals and the French Revolution*. On the Jacobins' egalitarian reforms see Gross, *Fair shares for all*.

12 Leith, *Space and revolution*.

people with a great number of images that they have revered for a long time, and which today are a source of religious errors'. As part of the nation's regeneration visions of ignorance had to be replaced with the reality of reason and priestly prestige had to be usurped by the truth of nature. Following a Lockean epistemology, Fabre d'Eglantine reasoned that even images provided the basis for knowledge and that the working of memory depended absolutely upon them. Consequently, the use of appropriate images would quickly engrave the calendar in the memory of the people.[13] As he explained to the Convention, the new calendar sought to divert the people's imagination from 'fantastic objects' and towards the more practical reality of agriculture and the four seasons. By such synecdochical methods, the calendar would become part of the fabric of everyday life and help to cultivate the spirit of the French people, ultimately forming the structure for a festive year that would replace the Catholic liturgical cycle with the purified, rational symbolism of the republic. The calendar, its proponents argued, would enact a cultural revolution as it marked the earth's passage around the sun.

The metrological and agriculturally-based names of the calendar not only made it more accessible to the peasantry, easier to understand and less demanding to memorise, but they also added to what might be termed the calendar's cultural capital, emphasising its links with classical civilisation and the natural order. As such, the committee of public instruction sought to avoid political controversy by providing a calendar for the nation as a whole, rather than commemorating individuals or political faction.[14] Links with agriculture and the seasons, while also having their own political associations, escaped the charge of divisiveness and offered France a bucolic vision of a nation in harmony with itself and the natural world. Recognising this common dream of a regenerated France, the Convention applauded Fabre d'Eglantine's report describing the new nomenclature, at least according to the pro-government newspaper, the *Moniteur*.[15]

Annuaire du cultivateur

Romme had already demonstrated his concern with the land and agricultural reform as a writer and editor of the journals *Feuille Villageoise* and *Feuille du Cultivateur*. In part inspired by his time as a gentleman-farmer in Riom in the

13 'rempli la mémoire du peuple d'un nombre considérable d'images qu'il a longtemps révérées, et qui sont encore aujourd'hui la source de ses erreurs religieuses': Fabre d'Eglantine, *Rapport*, 24 Oct. 1793.

14 Strictly speaking, Fabre d'Eglantine was reporting in the name of the Commission formed for the nomenclature of months and days: *Procès-verbaux*, 695 n. 3.

15 *Le Moniteur*, 'du 5 du second mois', 144.

Puy-de-Dôme, these publications sought to inform their readers of the most scientific and modern agricultural reforms, and he received advice from such noted agronomists and horticulturists as André Thouin, Louis-Jean-Marie Daubenton and the 'tuber impresario' Antoine-Augustin Parmentier (the agronomist responsible for the park at Enghien as well as somewhat quixotic attempts to encourage reluctant farmers to plant potatoes).[16] Some 15,000 subscribed to the *Feuille Villageois*, but, Romme realised, more could be reached though an almanac, one of the most popular forms of publishing, offering a useful compendium of facts and seasonal advice, as well as agronomic theorising. Together with Dubois and Lefebvre, who had also written for the *Feuille du Cultivateur*, Romme helped to publish the resulting *Annuaire du cultivateur*, an almanac for the new republican year, incorporating a simplified encyclopaedia of botany, agriculture and animal husbandry. A total of 2,000 copies was printed to be distributed to elected representatives and government officials. The intended audience was broader than this official group, and the annual was to be printed in the chef-lieu of each department and then distributed to each commune, under the watchful eye of officials, in order fulfil the Jacobin dream of widespread educational instruction.[17]

The *Annuaire* contained agricultural advice and information for each day of the year, dedicating each day in the manner of saints' days or Maréchal's *Almanach des honnetes-gens* to a different plant, animal or agricultural implement.[18] Each *décadi* was linked to an agricultural implement, such as the final day of the month of *floréal*, which praised the *houllette* (a form of crook used to remove troublesome stones from fields). The almanac suggested that readers could learn about these implements on their day of rest: instruction on the *houllette* was followed ten days later by tuition on the beneficial uses of the *pioche* (an agricultural pickaxe). The fifth day, the *quintidi*, which could be used as a day of rest by some officials and school children, was dedicated to a beast of some sort: for example, the first *quintidi* of the year was devoted to the horse, while the crayfish could be found on the last. The association with animals – often, with edible ones – suggested that the *quintidi* may have become some sort of republican 'high day', a less formal version of the *décadi*, when families might gather around the table.[19]

16 Garrone, *Gilbert Romme*, 335. We are in Colin Jones's debt for Parmentier's descriptive adjunct: *Great nation*, 352.

17 Romme, *Annuaire du cultivateur*, 1–2. On the educational aims of the Jacobins and the *Annuaire* see Gross, *Fair shares*, 188.

18 C. G. Romme, 'Annuaire du cultivateur', Museo del risorgimento, Milan, A2 Romme, 1794, and *Annuaire du cultivateur*.

19 The distinction between the ordinary days of the *décade* and the *décadi* or *quintidi* was also underlined in print: republican almanacs also tended to emphasis the *quintidi* and

The promise offered by such a calendar was one of a better educated peasantry, more able to exploit the land, and one which proposed a happier and better-fed future. Republican purity and wealth replaced royal corruption and death. The frontispiece to Millin's *Annuaire républicain* (fig. 4) made this contrast explicit, showing, as Simon Schama notes, the 'vanquishing of the old Gregorian tyrannies by the simplicity of rural husbandry', although the caveat that the 'simplicity of rural husbandry' was to be augmented by the application of agronomic science should be added.[20] However, this bucolic idyll was not one of rest, but rather an imagined space of virtuous labour, drawing on a long-developed technocratic desire to increase rural productivity and on the Revolutionaries' belief in the importance of work.[21] Citizens were not passive dwellers on the land, but actively contributed to the republic through their labours, particularly in the countryside. Again, republican language is telling. For example, Patrice Higonnet draws attention to the Jacobin '"politically correct" verbiage' that preferred the use of the word *cultivateur* rather than *paysan* for countryfolk, casting them as useful producers, rather than passive inhabitants of the land.[22] The agrarian idyll suggested by the calendar was one of productivity, rather than compliant consumption.

A vision of a rural democracy had informed American revolutionaries, notably John Dickinson, who argued that classical democracy could be inculcated not just in small city-states, but in the larger American colonies.[23] In was in these predominately agricultural states, where individuals could control their own destiny, that the ideal of virtuous citizen-farmer could take root. The idea of a regenerative relationship with the soil also found fertile ground among members of Europe's 'republic of letters'. French interest in the American democratic experiment helped to popularise works such as Crèvecoeur's *Letters from an American farmer* which considered the moral nature of the new nation and the idea of the republic. His work was widely disseminated in periodical form.[24] While such thought drew on the peculiar – and to the French mind, somewhat primitive, if Rousseauistic – characteristics

décadi by upper case typography. The system also proposed a secular circuit of self-improvement, in contrast to the 'superstition' of the system of saints days.

20 Schama, *Citizens*, 771.

21 Mainz, *L'Image du travail*.

22 Higonnet, *Goodness beyond virtue*, 113.

23 J. Dickinson, *Letters from a farmer in Pennsylvania: to the inhabitants of the British colonies*, Philadelphia 1774.

24 J. H. St John de Crèvecoeur, *Letters from an American farmer: describing certain provincial situations, manners, and customs*, London 1782, and *Lettres d'un cultivateur américain*, Paris 1784.

of the New World, the rebirth of the French nation allowed certain revealing parallels to be drawn.[25] If the French people could become virtuous, then democracy could flourish in a state formerly corrupted, according to much revolutionary rhetoric, by centuries of monarchical and superstitious oppression. This, in general terms, was to be the lesson of 'agrarian idealism'. A pastoral vision of moral regeneration entered the public sphere: Rousseau's *Le Devin du village*, for example, showed the shepherd Colin rejecting the corrupt pleasures of urban life for the democratic and natural simplicity of the countryside.[26]

Regenerative imagery

Ties with the land and the natural world also spoke of the Revolutionaries' broader regenerative purpose, breaking not with the 'whole of the past', but 'retying a broken thread', as Mona Ozouf writes, with 'Nature herself, in her primal purity'.[27] 'Jacobin man', Higonnet notes, 'realized his potential through work effected in fruitful cooperation with others and with nature'.[28] While female imagery may have been increasingly sidelined during the Revolutionary period, particularly by the Jacobins, the republic also, as Madelyn Gurtwirh suggests, flirted with embracing what she calls the 'goddess cult' of the female allegorical form.[29] Nature figured typically as a female figure, whose breasts symbolised regeneration, offering succour to the young nation, a trope that was easily understood within the cultural context of Enlightenment representations of moral reform. As Gurtwirh notes, 'Rousseau had established maternal nursing as the precondition of national regeneration in his *Emile*'.[30]

This nurturing, regenerative symbolism could take various forms, notably that which derived from classicism, which included Egyptian influences. As the *Encylopédie* made clear, France had a specific connection to Isis, who was

25 E. Durand, *Mirage in the west: a history of the French image of American society to 1815*, Princeton, NJ 1957.

26 J. Letzter and R. Adelson, *Women writing opera: creativity and controversy in the age of the French Revolution*, Berkeley 2001, 58–60. On the competing visions of 'agrarian idealism' see J. Livesey, 'Agrarian ideology and commercial republicanism in the French Revolution', *Past & Present* clvii (1997), 94–121.

27 The classic statement of the importance placed on 'regeneration' by post-revisionist historians is at Ozouf, 'Regeneration', and *Festivals and the French Revolution*, 34.

28 Higonnet, *Goodness beyond virtue*, 112.

29 M. Gutwirth, *The twilight of the goddesses: women and representation in the French Revolutionary era*, New Brunswick, NJ 1992, 362.

30 Ibid. 344.

Figure 2. 'La Fontaine de la Régénération', 1796–7
(British Museum, French XVIIIc)

allegedly worshiped by the Gauls, and who, via her Latinate name of Leucothea, had given the city of Paris its name. Artists such as Louis Germain provided current images of the goddess in the fashionable, if academically controversial, neo-classical mode.[31] Isis was understood to represent rebirth and plenty, and her worship by the ancient Egyptians in spring offered her as a suitable symbol for the regeneration of France.[32] Rather than suckling her son, Horus, Isis was shown as a giant *papier-mâché* statue of Regeneration in the Festival of Reunion on 10 August 1793, bringing new life to France after the strife of Federalist division (*see* figure 2) and commemorated on the 5 *décime* coin in the new, decimal currency (*see* figure 3).[33] Regeneration, such symbolism suggested, would bring wealth to France.

The calendar lent itself to being populated by such gods and goddesses. After all, the Gregorian calendar, had a long history of religious symbolism, and each month could be illustrated by a different figure. Large calendric engravings, such as the J. F. Lefevre's *Calendrier national* for 1792, could be

31 C. Jones, *Paris: biography of a city*, London 2004, 4; *Encyclopédie*.

32 Egypt also made its presence felt linguistically, as the terms *épagomènes* the technical terms used in the reports on the calendar to describe the *sans-culottides* recalled the festivals of the Egyptian year; the structure of the new calendar also remained indebted to the Egyptian system. See also Maréchal, *Almanach des honnêtes-gens*, and 'Calendrier', *Encyclopédie*, ii. 552–5.

33 Gutwirth, *Twilight of the goddesses*; Ozouf, *Festivals and the French Revolution*, 84.

Figure 3. 5 Décimes, 1793/Year II (BNF)
The 10 August 1793 was the date of the Festival of Reunion

decorated by pagan figures, including bare-chested females, and a prominent placing for a nursing mother, balancing the figure of justice on the right of the engraving (*see* figures 4 and 5).

A suite of scientific and historical charts and tables embellished the neatly-divided arch of the calendar (which dominates the engraving, foregrounding the calendar as a natural sequence of dates, rather than a religious construction), offering a national vision of the nation, bound by geography, population and secular time, rather than royal or religious cosmology. The calendar could be seen as republican diagram: as a recent historian of the calendar, Sanja Perovic notes, this 'new division of time was linked with a new sense of the French nation as defined by natural boundaries and no longer the body of the King'.[34] Once the republican calendar was introduced and the royalist, Catholic Gregorian calendar abolished, then the transition to a new symbolic order would be complete.

By drawing on such imagery, the calendar pointed to the productive capacity of the soil and the promise of the French nation, as well as promising the harmony and stability that Ozouf suggests was the main message of the statue and festival.[35] The use of nature as a symbol not only suggested peace

34 J. F. Lefevre, coloured etching on silk; *Calendrier national calculé pour 30 ans et présenté à las Convention nationale le 31 décembre 1792 … par le républicain J. F. Lefevre*, M. P. Vellet scripcit 1792, BNF, RC-A-10439; Perovic, 'Untameable time'.
35 Ozouf, *Festivals and the French Revolution*, 84.

Figure 4. Lefevre, 'Calendrier national', Paris [1792] (BNF)

Figure 5. Lefevre, 'Calendrier national' (detail)

and harmony, but also growth, production and work in the service of the nation. The names of the republican months and days reinforced this message, tying the new political order to a desire for a wealthier and more productive future.

Agriculture, of course, was seen to be central to the production of wealth, not least because of the influence of physiocratic thought.[36] While admitting that France was ripe for agricultural improvement and acknowledging the need to improve the connection between the land and virtue, the imagery and language surrounding the calendar appealed to a pastoral vision outside of time, suggesting that the revolution in politics would lead to a natural revolution in morals, restoring France and her people to the harmony and simplicity of the agricultural year, rather than the troubles and upheavals of the recent past. Louis Lafitte's depiction of Liberty – a representative of the nation – showed the nation in female form at peace.[37] It provided a vision of time which imagined the French nation as a whole, secure in a new, republican present, free from disorder, and – crucially – it placed this vision within a moral and economic context.

These rural references and allusions provide the central motif in perhaps the most well-known series of illustrations related to the calendar, the engravings made by Salvatore Tresca after Lafitte in Year II (*see* figure 6). It is also appears initially as an image of harmony, in contrast to the religious and political divisions that the calendar would cause. Liberty is shown as an often partially-robed female figure in a sequence of rural idylls, accompanied by a short poem based on the republican month.[38] The months of spring and summer were depicted as times of natural bounty, while autumn and winter also figured as productive periods: in *vendémiaire*, Liberty gathers the fruit of

36 E. Fox-Genovese, *The origins of Physiocracy: economic revolution and social order in eighteenth-century France*, Ithaca 1976; *Procès-verbaux*; Whatmore, *Republicanism and the French Revolution*.

37 On the contrast between male, heroic statues and the 'maternal peace' of many revolutionary depictions of nature or the Revolution see Ozouf, *Festivals and the French Revolution*, 206.

38 Louis Lafitte, 1770–1828. J. Duchesne, *Catalogue des tableaux, dessins, estampes, livres, médailles, coquilles et curiosités; planches gravées, curiosités, ouvrages à figures, livres sur les arts, littérature: du cabinet du feu m. Louis Lafitte, peintre, catalogue des tableaux*, Paris 1828, pp. iii–iv. The sequence has been exploited by numerous modern firms as advertising material, promoting everything from the cough medicine, Résyl, manufactured by Laboratoires Ciba, Lyon, printed by Léon Ullmann, Paris, to a vintage of Côte-de-Rhone, 'Vieux Clocher', Arnoux et Fils, Vacqueyras-Vaucluse, 1988, *Winegirl.ch*. http://www.winegirl.ch/Series/Calendar/summer.htm [29 Oct. 2005]. Tresca's engraving also appear in Baczko, 'Calendrier républicain', 70, and M. Vovelle, *La Révolution Francaise: images et récit, 1789–1799*, iv, Paris 1986, 219. A more detailed description of the paintings can be found at Mainz, *L'Image du travail*, 185–6.

Figure 6. Louis Lafitte, 'Messidor', engraved Salvatore Tresca, Paris 1793 (BNF)

the vine; during *brumaire* she gathers wood for winter and brings the sheep down from the hills, resembling in Valérie Mainz's words a 'profane Madonna'; then hunts or spins for the next two months. Finally, during *ventôse*, since the sun is in the sign of Pisces, she fishes.[39]

The symbolism displayed in the work is rich. Lafitte drew on a range of visual registers, mixing a neo-classical sensibility appropriate for the Revolutionary era with a rococo sensuousness, perhaps best seen in the drawing of lambs, flowers and doves.[40] Liberty-as-nymph is not shown as an overly stern, classical or stoical female (with the exception of the winter huntress), but a more sexual, sensuous being: a goat is seen to nibble at the vine leaves she gathers, and in *thermidor* she clasps a swan to her breast as she bathes (while a

39 Mainz, *L'Image du travail*. 186.

40 The range of Lafitte's artistic influences can be seen in the sale of his collection of art: Duchesne, *Catalogue des tableaux*.

brass figure provocatively mounts a jug to the right of the frame).[41] Each image also refers to the appropriate astrological sign for the month, drawn faintly at the top left or right of every frame, and the astrological setting usually forms part of the poetic allegory printed beneath each scene. While the imagery is in many ways universal, avoiding reference to contemporary place or time, the text that surrounds it subtly locates the calendar, itself representing an abstract ideal of an unchanging cycle of months and seasons, within the specifics of contemporary time and space. Brief tables of astronomical data detail the rising and setting of the sun and the length of the day in Paris, Rome, Madrid, London, St Petersburg and Vienna, providing a geographical and chronological reference point, and a reminder of the daily, rather than annual, scale of time. Each month's equivalent in the Gregorian Calendar is listed below the republican month, followed by the dominant astrological sign. The artist and engraver include their names in the customary places under the picture and the publisher includes his address at the foot of the page along with the note 'deposited at the Bibliothéque Nationale'; such devices were perfectly usual for printed materials, but also served to place the calendar in time and place, subtly reminding the reader or observer of the constructed nature of the calendar, of its function as a printed document as much as an abstract or ideal conception of the year.[42] While Lafitte and Tresca's engravings of the calendar were aimed at the luxury end of the market and were to be viewed in a particular setting – such as the study of the collector of prints, other calendars, almanacs and annuals had a more prosaic function, being sited on walls, desks or carried about the person for ready-reference.[43]

With the exception of the text noting that the printed calendar had been deposited in the Bibliothéque nationale, there was little to remind its audience of the national, republican nature of the calendar. Instead of connecting the year to the recent history of France and the Revolution, whether by showing military victories or Parisian *journées*, Tresca and Lafitte's depiction

41 The lines for *thermidor* read 'Sous un soleil brûlant l'Eau qui tombe en cascade/& les Jeux séduisants de ce Signe amoureux/Aux délices du Bain, invitent la Navade/Qui dans l'onde limpide attiédira ses feux.' On the tensions between sensuality and sensibility see L. Walsh, '"Arms to be kissed a thousand times": reservations about lust in Diderot's art criticism', in G. Perry and M. Rossington (eds), *Femininity and masculinity in eighteenth-century art and culture*, Manchester 1994, 162–83.

42 On the complex relationship between citizen and print culture during the Revolution see R. Taws, 'Circulation and authority: print culture and the fabrication of political identity in revolutionary Paris', unpubl. PhD diss, London 2005.

43 On the continuation of the luxury market for printed materials during the Revolution see A. Griffiths, *Prints for books: book illustration in France, 1760–1800*, London 2004.

of the calendar proposed a harmonious year, sharing the universal values or concepts of antiquity and the bounty of the natural world. While the calendar was republican in the sense that it avoids any monarchical (or Christian) reference, it does not appear to be revolutionary in any obvious sense, in that it makes no reference to the origins of the republic or to recent political events, leaders or groups. Yet the choice of a classical, pastoral trope can also be seen as an attempt to represent the republic in a particular mode, suggesting a happy vision of the future of France opened up by the gains and ideals of the Revolution. Instead of the 'superstition', subservience and corruption of the monarchical and Catholic year depicted in the annual *Almanach royal* (which combined a list of offices, royal and courtly events, religious festivals and dominical days, and the frontispiece of which depicted the royal figure of the king dominating the small scrolls of the calendar), the republican calendar proposed an agrarian idyll, in which the seasons and the land were in harmony with the citizenry.[44] The use of female imagery – the series of depictions of Liberty – again suggested harmony and balance, a nation undivided.

Lafitte's series of illustrations and Tresca's engravings point to cultural and political complexities of the calendar, ranging from assumptions about its materiality and use to the cultural associations of its imagery and symbolism. Visual and textual links between agricultural production and the natural world were intended to cultivate use of the calendar and inform a particular vision of the republic. From the names of the months to the imagery used by Lafitte, the republican calendar suggested a connection between the republic and the natural world: the constitution that the calendar commemorated provided a shortcut to the sublime. This aesthetic ideal of the pastoral had obvious ideological implications, developed by many of the leading thinkers of the day. The land, the seasons, and the celestial actions of the earth, sun, moon and stars were not simply observable or useful facts of everyday life, but informed a moral and political agenda. In particular, a harmonious relationship with the land could lead to an increase in citizens' virtue, allowing republicanism and democracy to flourish at the expense of corruption and despotism. Agriculture, as Saint-Just wrote, was the 'mother of [good] *moeurs*'.[45] A productive relationship with the land created virtue; despite anodyne appearances, the imagery of Tresca and Lafitte's calendar contained

44 *Almanach royal pour l'année*, Paris 1700. Competitors included the *Almanach de gotha: contenant diverses connaissances curieuses et utiles pour l'année*, Gotha 1763. See S. Perovic, 'Between the volcano and the sun: the Revolutionary Calendar and Sylvain Maréchal', unpubl. paper, Modern France Workshop, University of Chicago, 25 Feb. 2005.
45 Saint-Just, 'La Mère des mœurs', quoted in Higonnet, *Goodness beyond virtue*, 115.

a coded restatement of republican ideals.[46] Inspired by a Rousseauean vision of nature and an ingrained respect for the agrarian republicanism of Cincinnatus, who left his plough to defend the Roman republic, Jacobins, as Patrice Higonnet notes, had a 'strong prejudice for the beloved eighteenth-century fiction of the republican yeoman farmer of Jeffersonian fame'.[47] The bounty of nature allowed rest and the promise of a peaceful, well-fed future, but its agricultural setting and classical aesthetic also stressed the importance of work.[48]

Symbolism, of course, could also operate by calling to mind elisions, and the calendar could not but remind its viewer or reader of the system that it had replaced. The absence of the Gregorian Calendar was made known by its use for the dating of zodiacal events (such as 20/21 July, the sun corresponding with Libra); the novelty of the calendar underscored the republic's newness and perhaps worked to suggest that the republic was an alternative to other systems of government, rather than the only, inevitable form of sovereignty. Nor was the notion of agrarian idealism uncontested. Republicans continued to argue over whether land should be redistributed equally or whether the status of the productive class had to be maintained: as James Livesey notes, 'an agrarian ideal in the Revolution should not obscure the differences between different versions of that agrarian vision'.[49]

Such potential conflicts are present to a greater or lesser degree in other cultural effusions of the calendar. While Romme laboured on his *Almanach des cultivateurs*, editing copy and attempting to keep the *Feuille des Cultivateurs* financially afloat, the Jacobin Paris theatre responded swiftly to the new calendar. A one-act play, *L'Heureuse Décade*, was performed in the Théâtre du Vaudeville on 5 *brumaire* II (26 October 1793), subtitled a 'patriotic amusement'. The play was set in a village square while the *décadi* was being celebrated as a day of rest. The civic day was contrasted with Sundays and feast days which, as the Jacobin characters declaimed, undermined the efforts being made in the name of the republic: as the main character, Père Socle, sang, 'Days of religious festivals/ are lost to the country'.[50] Such patriotic and

46 Whatmore, *Republicanism and the French Revolution*, 97.

47 Higonnet, *Goodness beyond virtue*, 115, 138. On the complexities of Jacobin economic and political thought see also Gross, *Fair shares*, and on the Directorial period, Livesey, *Making democracy*.

48 Whatmore, *Republicanism and the French Revolution*, 97.

49 Livesey, 'Agrarian ideology', 104.

50 'Les jours aux fêtes consacrés,/ Etaient perdus pour la patrie': M. Barré, F. P. Auguste Leger and J. R. Rosières, *L'Heureuse Décade, divertissement patriotique, en un acte et en vaudevilles, des citoyens barré, léger et rosières*, Paris 1793, repr. in *Répertoire du théâtre républicain: ou, recueil de pièces imprimées avant, pendant et après la république franeaise*, vii, Geneva 1986, scene 1.

revolutionary fervour had some public appeal and the published play went into at least two editions. A two-act sequel, entitled *Seconde Décade ou le double mariage*, followed on the 4 *nivôse* II (24 December 1793), at the Théâtre Patriotique. Père Socle continued to offer advice on republican idealism to his now newly-married son, balancing agricultural production and service to the republic: 'Undoubtedly, to be in one's youth a soldier and ploughman, is to render double service to one's country, and gloriously fulfils the task of a republican.'[51] The comparisons with Cincinnatus are clear. Plays also drew attention to the problems with the calendar: *L'Intérieur d'un ménage républicain*, performed in the Théâtre de l'Opéra-comique on 15 *nivôse* II (4 January 1794) contained a scene discussing the abolition of religious festivals and Sundays (the day being 'the day of Toussaint! One of the biggest festivals of the year!').[52] Amelie, the sister of the young republican, sang with regret: 'It is another change/ That will cause you more sorrow/ The week has been lengthened/ Without due care.'[53]

The market for such patriotic, republican entertainments outside of Paris was more limited, but the connections between the calendar and aspects of French commercial and working life were still made. Bordeaux, under close Jacobin control following the Federalist revolt of the summer and surrounded by a countryside populated with peasants largely suspicious of innovations imported from the capital, was treated to a more didactic, if allusive, theatrical event.[54] During the *sans-culottides* of the Year II (17–21 September 1793), Marc-Antoine Jullien 'de Paris', representing the committee of public safety, produced a ballet at the city's Grand Théâtre entitled 'Le Calendrier républicain'.[55] The ballet consisted of a series of mostly female personifications of each month of the republican year, who paraded on stage in front of a temple, accompanied by a suitably dressed retinue. The month of *vendémiaire* appeared first, wearing a pair of flesh-tone trousers, draped with vines and carrying grapes, as did the grape-pickers who followed her on stage. *Brumaire*

51 'Sans doute, être dans sa jeunesse soldat & laboureur, c'est se rendre doublement utile à sa patrie, & remplir glorieusement la tâche de républicain': A. P. Bellement, *La Seconde Décade, ou le Double Mariage, opéra patriotique en 2 actes et en vaudeville, suite de l'"Heureuse Décade"*, Paris 4 nivôse II [1793].

52 'le jour de la Toussaint! Une des plus grandes fêtes de l'année'.

53 'un autre changement/ Qui te causera plus de peine,/ C'est que sans nul ménagement,/ L'on vient d'allonger la semaine'.

54 A. I. Forrest, *Society and politics in revolutionary Bordeaux*, Oxford 1975; Eric Saugera, *Bordeaux, port négrier: chronologie, économie, idéologie, XVIIe–XIXe siècles*, Biarritz 1995.

55 Marc-Antoine Julien 'de Paris', b. 1744, d. 1821. The connection between time and the arts appeared to continue to interest Jullien who fled France after accusations of involvement in the Babeuf plot: M.-A. Jullien, *Essai sur l'emploi du temps, ou méthode qui a pour objet de bien régler l'emploi du temps*, Paris 1810.

followed, dressed in grey gauze and accompanied by four similarly dressed children portraying the gathering of storm-clouds and dark mists. The chill of *frimaire* was heralded by a figure dressed in the skins of wild animals and carrying a bow and arrow, while *nivôse* was dressed as white as snow, holding a trivet with a flame achieved by means of inflammable ethyl alcohol. An urn-carrying water-nymph accompanied by an entourage of umbrella-carrying old men and women personified watery *pluviôse*. *Ventôse* and her retinue dressed as winds, while *germinal* wore flowers, with a blossoming cherry branch in hand. *Floréal*, dressed as Flora, arrived amidst 'small zephyrs' and garlands of flowers. Garbed entirely in green and with a belt of violets, she was accompanied by four children bearing watering pots. The figure of *messidor* was crowned with ears of corn. *Thermidor*, a male figure, entered almost nude, wearing a large sun on his chest, his face covered with sweat and holding flaming torches in his hands. Four peasants wiped his brow. Finally, *fructidor* arrived bearing a horn of plenty, pursued by the five festivals of the *sans-culottides*, dressed in costumes representing Genius, Virtue, Work, Opinion and Recompense and succeeded by a troupe of young girls carrying a small cardboard model of the Paris Pantheon.

More allegorical republican figures followed: Liberty (carried by four *sans-culottes*), then Equality (carried by a ploughman, a rich man, a Moor and a mulatto). Fraternity (and a form of racial equality) was represented by three women, one of whom was black, and all of whom shared the same coat. Surveillance appeared with an all-seeing eye on his front, a triangle on his chest, all topped by a cloak decorated with eyes. Finally, Victory and Reason entered carrying a list of the Rights of Man. The months then arranged themselves as the Four Seasons and began their dance. The performance finished with the 'adoption' of a law, a civic hymn, and a celebration of marriage and eulogy to the production of children.[56] The ballet served as an ideal model for the proposed *fêtes décadaires* and *nationales* that were to take place in every commune. Few could provide either the artistic skills or afford the cost of such theatrical emulation, but similar imagery of pastoral abundance and the call for the birth of republican children was repeated in the following year's festival of the supreme being in Paris, when a female figure carrying sheaves of wheat together with a fruit tree appeared in a chariot led by bullocks.[57]

56 J. Brouillard, 'Un Ballet original', *Revue historique de Bordeaux et du département de la Gironde* v (1912), 277–8.

57 Criticism of the cost of festivals and many other aspects of the calendar can be found in post-Thermidorian pamphlets, such as C. Saint-Aubin, *Réflexions sur le nouveau calendrier et sur la nécessité d'élaguer au moins les décadis de cet avorton du système décimal; avec un mot en passant sur les effets du fanatisme politique comparés à ceux du fanatisme religieux*, [n.p.] 1795,

Unlike the plays performed in Paris, the Bordeaux ballet must have cut close to the bone. Jullien, who was relatively close to Robespierre (he was imprisoned after the anti-Jacobin coup of Thermidor), represented the committee of public instruction's attempt to regain control of the Federalist city in the heart of the Gironde. The ballet vividly demonstrated the Jacobin belief, expressed by Fabre d'Eglantine, in the power of images and propaganda to influence the citizenry; Jullien also included the rigour of Surveillance as a reminder of the power of the virtuous citizenry to root out the bad. It provided an alternative vision of the nation, shorn of monarchical or Christian imagery to represent the state or its people, and instead offered nature as a model for the nation. More pointedly, the ballet underscored the Jacobin commitment to equality, and included racial equality, by showing black and white in fraternal harmony; a controversial point in a city which for so many years had depended for her wealth on the Caribbean coffee, sugar and slave trade.

Although these figures may have been taken to undermine the commercial basis for the city, slavery had been abolished only in France and would continue in the colonies until 4 February 1794 (although, of course, the revolt in Saint Dominque had effectively emancipated many slaves before this). The rest of the ballet underlined the relationship between the new nation and production, whether of children, of families, or – as in *messidor* – in agricultural terms. Republican *moeurs* were shown in the production of children; as Robert Darnton suggests, 'virtue was virile', reproduction was a 'civic virtue' and 'bachelors were excoriated as non-patriotic'.[58] Furthermore, work made its appearance among the festivals of the *sans-culottides* as the Festival of Labour.

At this time, and somewhat to the north of Bordeaux, Jullien's colleague, Jean-Baptiste-Jérôme Bô, sent on mission by the Convention to the heart of the counter-revolution in Nantes, presented a similar, but more explicit message about the republic, its calendar and the virtue of labour. He railed against the problems caused in the countryside by the obstinate use of the Gregorian Calendar and announced to the local population that superstition, which had caused so much trouble in the Vendée, still had its hold, while observance of religious feasts and Sundays led to idleness. Furthermore, all good citizens had a duty to work regardless of the old calendar: 'When your rest is excessive and harmful to the common interest [then] you are bad

12. See G. Taylor, *The French Revolution and the London stage, 1789–1805*, Cambridge 2000, 82.

58 R. Darnton, 'What was revolutionary about the French Revolution?', *New York Review of Books*, 19 Jan. 1989, 3–10; V. M. Moghadam, 'Gender and revolutions', in J. Foran (ed.), *Theorizing revolutions*, London 1997, 144.

citizens [...] The working man has need of rest, and your legislators have dedicated to rest the tenth day of each *décadi* [otherwise] work is a necessity and a virtue, and rest would be a crime.' Bô exhorted the populace to work hard, and also to attend the Temple of Reason on the *décadi* in order to gain republican 'strength' and to avoid vice.[59] In the geographic heart of the counter-revolution, the republican calendar could not be depicted in terms of nature and harmony alone, but required more strident tones, arguing that laziness was a crime. The republic required citizens' labour. Clearly, the use of one's time was associated with the cause of the republic and the state could make a legitimate, and increasing, claim on citizens' time. Off the stage, two merchants, Jacques Largetau and Duray Longua, from Libourne in the Gironde, fell foul of this feeling and were accused of selfishness. They had never attended the meetings of the *sociétés populaires* or their sections and contributed nothing to the republic. The *commission militaire* fined them 10,000 *livres*. The *Courrier républicain* approved of the *jugement* on the two men who 'dedicated their time only to financial speculation and to have made no sacrifice for the republic'.[60] In the new republic, as Valérie Mainz has suggested, citizens had to provide good account for their time; indeed the image of self-sacrifice and happy labour in the aid of the state were common strands in the iconography of the period.[61] But the shift from performing the calendar on stage, to enacting it in everyday life would be problematic at best. While Revolutionaries could provide republican images, symbols and performances, the transition from stage or page to the transformation in daily, internalised actions was a more difficult process. Mere observance of republican or revolutionary symbolism could not be relied upon to inculcate moral regeneration, despite Fabre d'Eglantine's belief in the efficacy of this process. Perhaps transformation could take place elsewhere, notably in the personal, daily negotiations made between self, the wider word and perhaps even the nation, effected by clock and calendar.

The materiality of the new time

The new form of time presupposed a new form of engagement between the citizen and the state, suggesting a very personal link between the personal measure of time, and the wider context in which it was framed. In 1794 *Les*

59 J.-B.-J. Bô, 'Le Représentant du peuple à nantes, près de l'armée de l'ouest, & dans les départements en dépendants, aux habitans des campagnes', 4 messidor II (22 June 1794), AD Loire-Atlantique.

60 *Courrier républicain*, 21 *brumaire* II (11 Nov. 1793).

61 Mainz, *L'Image du travail*, p. xx.

Décades républicaines carried an advertisement for a 'montre économique', priced at fifteen *sols*. The advertisement claimed that the device boasted both utility and patriotism, allowing the purchaser to make use of the new, republican hours:

> For the Sans-Culottes, an indicating dial, with which one can very easily [...] comprehend the decimal hours, by the use of an unmodified clock. One only needs to see what time it is on another watch, and then to take the same point on this dial and follow the dotted line from that point, and which ends up on the decimal hour and minute. One will have the exact correspondence between the new hours and the old.[62]

Such devices no doubt fulfilled various functions, from being a novelty to those interested in such things or a practical solution to recording the time for those in official positions, to a means of demonstrating one's republicanism for those who wished to do so. A similar dial was also marketed in Germany, for example where expressions of support for Jacobinism took on specific political meanings.[63] Perhaps unlike the experience of watching a ballet, the observance of the republican calendar and, in particular, the use of decimal time, was a far more active and personal engagement with the new culture. The process of comparison was an active one, a procedure involving physical as well as mental activity.

Concordances, whether calendrical or, as in the case Periaux's *Tableau du concordance entre les divisions décimales et les divisions anciennes* (1794) (*see* figure 7), which translated the old hours into the new (or vice-versa), engaged the citizen in a process of comparison, at once underscoring the difference between the republican *now* and the monarchical *then* and displaying time as a neatly laid out table, abstract and free of undue meaning.[64] The use of the decimal time or a republican almanac, translating the old dates into the new, was a process very much like that conceived by Fabre d'Eglantine, one in which republican objects replaced those of religious 'fantasy'. The same process of translation of dates took place on the masthead of every paper and journal, reinforcing, if Benedict Anderson is correct, the increasingly important idea of the nation.[65]

62 A. Sérieys, *Introduction aux décades républicaines, ou à l'histoire abrégée de la république française, etc.*, Paris 1794.

63 German *cadran comparatif* for the Year III, repr. in S. Seifert, *Die Zeit schlägt ein neues Buch in der Geschichte auf: zum Französischen Revolutionskalender und zu seiner aufnahme in Deutschland*, Weimar 1989.

64 P. Periaux, *Tableau du concordance entre les divisions décimales et les divisions anciennes*, [Paris] 1794.

65 Anderson, *Imagined communities*, 24.

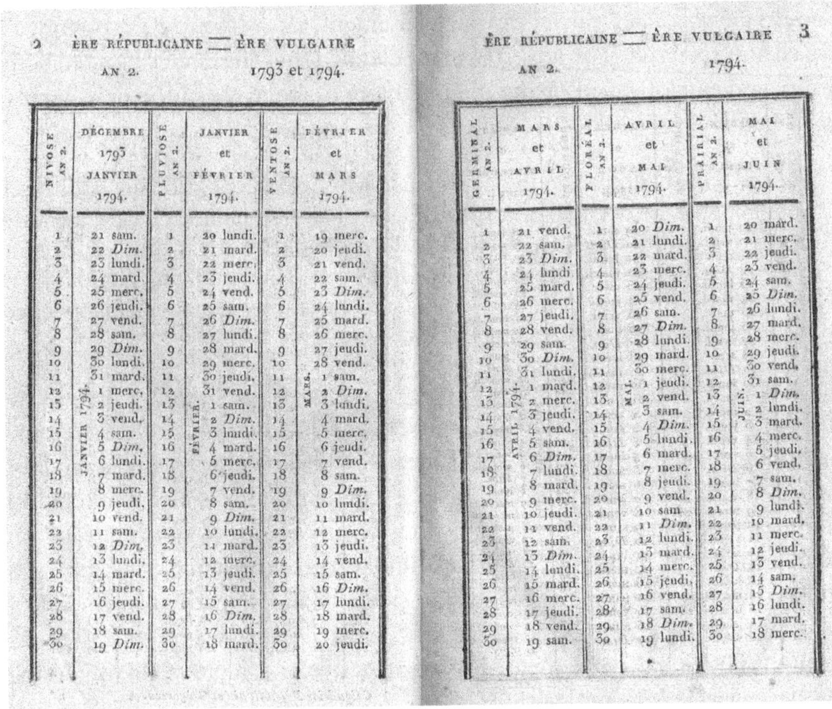

2 ÈRE RÉPUBLICAINE = ÈRE VULGAIRE

AN 2. 1793 et 1794.

NIVOSE AN 2.	DÉCEMBRE 1793 JANVIER 1794.	PLUVIOSE AN 2.	JANVIER et FÉVRIER 1794.	VENTOSE AN 2.	FÉVRIER et MARS 1794.
1	21 sam.	1	20 lundi.	1	19 merc.
2	22 *Dim.*	2	21 mard.	2	20 jeudi.
3	23 lundi.	3	22 merc.	3	21 vend.
4	24 mard.	4	23 jeudi.	4	22 sam.
5	25 merc.	5	24 vend.	5	23 *Dim.*
6	26 jeudi.	6	25 sam.	6	24 lundi.
7	27 vend.	7	26 *Dim.*	7	25 mard.
8	28 sam.	8	27 lundi.	8	26 merc.
9	29 *Dim.*	9	28 mard.	9	27 jeudi.
10	30 lundi.	10	29 merc.	10	28 vend.
11	31 mard.	11	30 jeudi.	11	MARS. 1 sam.
12	JANVIER 1794. 1 merc.	12	31 vend.	12	2 *Dim.*
13	2 jeudi.	13	FÉVRIER. 1 sam.	13	3 lundi.
14	3 vend.	14	2 *Dim.*	14	4 mard.
15	4 sam.	15	3 lundi.	15	5 merc.
16	5 *Dim.*	16	4 mard.	16	6 jeudi.
17	6 lundi.	17	5 merc.	17	7 vend.
18	7 mard.	18	6 jeudi.	18	8 sam.
19	8 merc.	19	7 vend.	19	9 *Dim.*
20	9 jeudi.	20	8 sam.	20	10 lundi.
21	10 vend.	21	9 *Dim.*	21	11 mard.
22	11 sam.	22	10 lundi.	22	12 merc.
23	12 *Dim.*	23	11 mard.	23	13 jeudi.
24	13 lundi.	24	12 merc.	24	14 vend.
25	14 mard.	25	13 jeudi.	25	15 sam.
26	15 merc.	26	14 vend.	26	16 *Dim.*
27	16 jeudi.	27	15 sam.	27	17 lundi.
28	17 vend.	28	16 *Dim.*	28	18 mard.
29	18 sam.	29	17 lundi.	29	19 merc.
30	19 *Dim.*	30	18 mard.	30	20 jeudi.

ÈRE RÉPUBLICAINE = ÈRE VULGAIRE 3

AN 2. 1794.

GERMINAL AN 2.	MARS et AVRIL 1794.	FLORÉAL AN 2.	AVRIL et MAI 1794.	PRAIRIAL AN 2.	MAI et JUIN 1794.
1	21 vend.	1	20 *Dim.*	1	20 mard.
2	22 sam.	2	21 lundi.	2	21 merc.
3	23 *Dim.*	3	22 mard.	3	22 jeudi.
4	24 lundi	4	23 merc.	4	23 vend.
5	25 mard.	5	24 jeudi.	5	24 sam.
6	26 merc.	6	25 vend.	6	25 *Dim.*
7	27 jeudi.	7	26 sam.	7	26 lundi.
8	28 vend.	8	27 *Dim.*	8	27 mard.
9	29 sam.	9	28 lundi.	9	28 merc.
10	30 *Dim.*	10	29 mard.	10	29 jeudi.
11	31 lundi.	11	30 merc.	11	30 vend.
12	AVRIL 1794. 1 mard.	12	MAI. 1 jeudi.	12	31 sam.
13	2 merc.	13	2 vend.	13	JUIN. 1 *Dim.*
14	3 jeudi.	14	3 sam.	14	2 lundi.
15	4 vend.	15	4 *Dim.*	15	3 mard.
16	5 sam.	16	5 lundi.	16	4 merc.
17	6 *Dim.*	17	6 mard.	17	5 jeudi.
18	7 lundi.	18	7 merc.	18	6 vend.
19	8 mard.	19	8 jeudi.	19	7 sam.
20	9 merc.	20	9 vend.	20	8 *Dim.*
21	10 jeudi.	21	10 sam.	21	9 lundi.
22	11 vend.	22	11 *Dim.*	22	10 mard.
23	12 sam.	23	12 lundi.	23	11 merc.
24	13 *Dim.*	24	13 mard.	24	12 jeudi.
25	14 lundi.	25	14 merc.	25	13 vend.
26	15 mard.	26	15 jeudi.	26	14 sam.
27	16 merc.	27	16 vend.	27	15 *Dim.*
28	17 jeudi.	28	17 sam.	28	16 lundi.
29	18 vend.	29	18 *Dim.*	29	17 mard.
30	19 sam.	30	19 lundi.	30	18 merc.

Figure 7. Periaux, *Concordance des calendriers republicain et grégorien*, Paris 1805

Despite this, the culture of the new regime was often still-born – often a novelty at best. Many of the examples in the preceding pages reveal an attempt to impose new cultural forms, rather than a response to more widely-held urges to describe a 'new world'. The Jacobin imagery of the calendar, according to Michel Vovelle, tended to express a 'bourgeois', urban conception of the bucolic ideal, far removed from the harsh realities of working the land. Tresca's engravings of the months of the republican year were, after all, deluxe items, with a price tag to match.[66] New watches and clocks were expensive, although republican time possibly created an emulatory, aspirational market among *sans-culottes* who could not afford a decimal watch, but could perhaps purchase the paper *montre économique*. The ballet offered a somewhat staid, academic performance, lacking in (intentional) humour, or obvious contemporary reference. The calendar was imbued not with the accretions of centuries of tradition, but the values of a

66 Vovelle, *La Révolution Francaise*, iv. 219. On the bourgeois nature of pre-revolutionary pastoral art see, for example, Letzter and Adelson, *Women writing opera*, 58.

well-educated, urban elite and its pastoral symbolism was that of a slightly prurient, literate class. In Richard Cobb's words, the new calendar

> was a carefully constructed edifice of fantasy and illusion. Even the new months – however pretty-sounding – had been the creations of poets who had never ploughed nor sown, who knew nothing of the burdens of the agricultural calendar, and who could in all comfort evoke the rigours of Nivôse, because they had fires laid for them at home, and travelled, well wrapped-up, in closed carriages, or who, in Brumaire, had torch-bearers walk before them.[67]

By the summer of 1793, as Romme and his scientific cohort worked on the calendar, the revolutionary festival had also shifted from the fecundity of earlier festivals and had, as Ozouf notes, 'become entirely what Edmund Leach has called a "formality"'.[68] Perhaps the same could be said of republican culture more broadly; Gurtwith suggests that this is true in the case of female symbolism, arguing that her nurturing form was replaced by the male or, later, reworked as the base representation of the Jacobin hag.[69]

While the grand claims of the calendar may have met with the failure described by Richard Cobb, its humbler implications tapped into a wider web of cultural and economic changes.[70] While the calendar's overarching symbolic system located it within a framework of the natural world, its material form placed it more prosaically within the world of work, print culture and consumption. Its potential for measuring the hours of work and days of labour must also be seen alongside the calendar as a cultural artefact that had to be manufactured, distributed, purchased and consumed; while the relationship between the calendar and work-discipline in a modern, capitalist sense, is problematic, the calendar can more properly be seen as an element of the cultural matrix that allowed the nation to be imagined through print or performance and, especially, through the object.[71] It was also a cultural matrix that relied on commerce to support itself, from printer to peddler, illustrator to bookseller. While the republican calendar was a state-supported institution, funded by the state's coffers and growing band of Jacobin activists, administrators and even revolutionary armies, entrepreneurs began a range of eighteenth-century 'public/private partnerships', winning printing contracts, marketing timepieces and sundials, and selling a plethora of repub-

67 R. Cobb, 'Thermidor or the retreat from fantasy,' *Encounter* lvii (1981), 31.

68 Ozouf, *Festivals and the French Revolution*, 84.

69 Gutwirth, *Twilight of the goddesses*.

70 On the political implications of consumption see C. Jones, 'The great chain of buying: medical advertisement, the bourgeois public sphere, and the origins of the French Revolution', *AHR* ci (1996), 13–40.

71 Spang, 'Paradigms and paranoia'.

lican almanacs or other republican publications, such as the *Catéchisme historique et révolutionnaire*, which offered itself to towns without newspapers, as it would provide a record of all events for the coming year.[72] Whether through ideological sympathies or financial necessity (after all, almanacs formed the commercial bedrock of many print shops), the new calendar encouraged a wealth of new almanacs, books of poetry, illustrated calendars and, of course, counter-revolutionary almanacs of various kinds. Government printers, the Bureau de Longitudes (which produced the *Connaissance des temps*, that vital instrument of navigation and trade), manufacturers of paper, wooden or ceramic sundials, as well as watch and clockmakers, playwrights and pamphleteers could all set about manufacturing new commentaries or measures of time.[73] While perhaps naive in their belief in the transformative power of images and symbols, Romme, Fabre d'Eglantine and the other members of the committee of public instruction were surely correct to place such store on the calendar's cultural context; surely the calendar's greatest impression on mental life was to be its connections to a wider web of cultural production, rather than grand, symbolic creations. Where these symbols did have the most influence, however, was in provoking opposition to the reform and the republic, most crucially over the question of religion; the flipside to the republican coin.

72 'Enfin, les habitants des villes et des campagnes qui n'ont point l'habitude des papiers publics, comme ceux qui les voient périodiquement, auront la satisfaction de trouver en petit volume un recueil exact de tout ce qui sera arrivé de plus remarquable dans le cours d'une année entière': *Catéchisme historique et révolutionnaire, contenant un recueil exact, authentique, chronologique et impartial des événements remarquables, ainsi que des actions d'éclat, traits de courage, d'héroïsme et autres qui sont arrivés, jour par jour*, [n.p.] [n.d.], 6–7. For a listing of some revolutionary almanacs see M. Sonnet, 'Les Almanachs politiques parus pendant la Révolution Francaise', *Bulletin des Bibliothèques de France* xxv (1980), 8.

73 On the relationship between print capitalism and nationalism see Anderson, *Imagined communities*, and D. A. Bell, *The cult of the nation in France: inventing nationalism, 1680–1800*, Cambridge, MA. 2001. On the *Connaissance du Temps* see Alder, *Measure of all things*, 24.

4

The Clash with Religion

M. Dimanche et Citoyen Décadi

One of the more popular humorous causeries of the Revolutionary period consisted of a dialogue between 'Monsieur Dimanche and Citoyen Décadi', two caricatures of the ideals of the Revolution and its antithesis. As Citizen Décadi noted, Sunday religious observance 'date[d] from the Deluge' and had, as M. Dimanche said, 'arrived hale and hearty down to the 18th Century'. It was contrasted with the rational but 'sterile' or 'ludicrous' reforms introduced by the republic. As the historian Alphonse Aulard observed, the contrast between these two embodied feast days formed a powerful metaphor for 'the quarrel between the Church and the state'. The clash between Messrs Sunday and Décadi has become a kind of shorthand for a complex dispute between the Revolutionary government and the Catholic authorities, one that was played out in the streets and churches of France, often in violent ways.[1] The usefulness of this metaphorical conceit extended beyond France. For example, the *émigré* journalist Peltier republished the dialogue in his London-based *Paris pendant l'année 1795*, and the *British Critic* found it of 'considerable humour'. Several London and provincial English papers published the exchange, as did continental papers, and the idea continued to be reworked in nineteenth-century periodicals such as the *Edinburgh Magazine*. Such personification powerfully distilled the conflict between the republic and religion, and the clash between Sundays and the tenth-day of the republican calendar is still found to be a useful motif for historians as a method of organising accounts of revolutionary religious policy.[2] While the calendar's nomenclature may have left its impression on the collective memory through the recycling of evocative terms such as *germinal* or *thermidor*

1 *Paris pendant l'année*, 1795, 6.

2 For example, C. W. Crawley (ed.), *New Cambridge modern history*, ix, Cambridge 1965, 175–6; Schama, *Citizens*, 775. See also H. Gough, *The Terror in the French Revolution*, London 1988, 47.

as the titles for left-leaning books and plays, its most indelible historical imprint may well be that of the republic's animosity towards the Catholic Church. Of course, contemporaries also made great play of the calendar's inherent anticlericalism: in 1806, when the calendar was finally abolished in favour of the Gregorian Calendar, many journals noted that it was a triumph for Catholic observance, suggesting that the calendar was seen in terms of the opposition between *laïcité* and Catholicism rather than that between Revolution and counter-revolution. Perhaps most famously, as the *abbé* Grégoire recalled in his memoires, Charles Gilbert Romme's aim was the destruction of Sunday. Grégoire responded: 'The day of Sunday,' I told him, 'existed before you; it will exist after you.'[3]

But whatever the reliability of the *abbé*'s memoirs, his recollections, like the distinction between M. Dimanche and Citizen Décadi, potentially mask the complexities of the relationship between the republic and religion and, in particular, the calendar's place within this fraught association. Indeed, the dialogue points to some of these intricacies. Citizen Décadi began the case for his defence by noting that the old calendar lost fifty-two days 'to industry' but that the 'decadary institution' only claimed thirty-six. The republican calendar was, he claimed, a 'religious and moral institution'. Sundays were rest days, and their disappearance carried important implications for work and industry, as much as for popular piety, and the new calendar also sought to promulgate moral sentiment just as the Gregorian Calendar had attempted to impart religious truths.

Despite the calendar's clear anticlerical legacy, it should not be assumed that the republic simply forged the calendar as a weapon for the dechristianiser's arsenal; rather, the calendar was intended to be a pillar of the regenerated state. It was constructive rather than destructive in its intent. While Sundays became a nexus for counter-revolutionary resistance and dechristianising campaigns, the wider aims of the calendar were to create what might be termed a civil religion, concerned with the moral regeneration of the citizen and, an admittedly fatally compromised, toleration for the Catholic faith. Of course, such intentions need to be qualified by the realities of the calendar's administration over its lifetime, but it is only with such ideals in mind that the attempt to reorganise time can be properly assessed, rather than dismissed simply as a 'preposterous and incredible' fabrication.[4] The calendar, like many republican institutions, also developed over time: it

3 Friguglietti, 'Gilbert Romme', 57; H. Grégoire, *Mémoires de Grégoire, ancien évêque de Blois, député à l'Assemblée Constituante et à la Convention Nationale, sénateur, membre de l'Institut*, Paris 1808.
4 *True Briton*, 1 Dec 1797.

is important to chart its shifting relationship to religion. It emerged from the mix of deism and secularism that characterised the French Enlightenment, it was introduced in a Convention alert to the language of what has come to be termed dechristianisation and, crucially, it was implemented by local administrators and representatives *en mission*, as they attempted to translate these philosophical discoveries into actual practice. This proved to be a difficult undertaking. For many villagers their local church meant far more than distant church hierarchies or republican governments. The church was often the centre of the community: it provided poor relief and education, and it was a meeting or rallying point at times of danger. Clerics were figures of authority, and often also operated as local 'fixers', helping to smooth out injustices or acting as a source of loans. The Church's pattern of festivals and rituals provided the bedrock for daily life and shape for the week and year. Even if many did not attend mass or take the sacraments, attacks on the Church were seen as attacks on the village community. Finally, during the Directory, calendar observance became an important plank in the administration's domestic policy, one that resulted from a long-running debate on the nature of the republic and how it was commemorated in its festivals.[5] The calendar was intended to be so much more than purely a means of suppressing religion.

The intent of the calendar's creators

Revolution and religion proved to be awkward bedfellows, despite a relatively promising start. While many of the First Estate had sided with the Third during the stand-off between the Crown and the national assembly in the early summer of 1789, with both claiming to support the interests of the nation, the alliance between the Revolutionaries and the clergy had long broken down by the hot summer of 1793. Successive governments consistently failed to incorporate the Catholic Church within the French state. The conflict between the sovereign will of the people and the Church's loyalty to the papacy and the universal Church had created an unavoidable fracture in the nation's body-politic, a split that had become most visible during the crisis of the Oath of the Constitution in 1790. In a symbolically loaded and politically fatal moment, the Constituent Assembly demanded that the newest, and largest, body of state employees – the clergymen of the newly-formed Constitutional Church – swear allegiance to the new constitution. Perhaps as many as two-thirds refused, and the mass of French clerics

5 Livesey, *Making democracy*.

joined their socially more elite brothers, bishops and abbots who had, with their noble connections, formed an important part of the great flight from France or who began to fear for their wellbeing, property and lives. Although attracting the support of the majority of the French population, who were used to a state-backed Church, and a considerable number of capable churchmen, the numbers of refractory priests pointed to future calamity facing not just the Church, but the nation.

In many ways the Civil Constitution led the national government into an increasingly fractious religious policy that made a religious settlement almost impossible. By the summer of 1793 the Constitutional Church – always, in Nigel Aston's words, an 'awkward amalgam of Christian and revolutionary fervour' – had largely lost its public footing. Denounced by Danton as an institution staffed by men who sought to establish 'their priestly throne on the ruins of liberty', and politically compromised by the association of many Constitutional ecclesiastics with Brissotin politics, the Church was in no position to resist the nascent dechristianisation movement; indeed, many of the newly appointed, Jacobin constitutional clergy gave it early succour. As Edmund Burke noted, with no small dose of rhetorical exaggeration, 'The Constitutional Clergy are not the Ministers of any religion; they are the agents and instruments of this horrible conspiracy against all morals.'[6] Thousands of clergymen abandoned their ministry, and for those that remained, little was left of their ecclesiastical role. Some republican *curés* found their pastoral calling more easily fulfilled in egalitarian street politics than by novel liturgies. Perhaps the most high-profile of such *curés rouges*, Jacques Roux, *vicaire* of Saint-Nicolas de Champs since 1791 and in a position of influence within the Cordeliers club, was able to place pressure on the committee of public safety to introduce the Maximum in 1793. More generally, in the popular or republican imagination the figure of supportive churchman, such as the archbishop of Paris, had been replaced by a shadowy cadre of *émigré* priests, encouraging counter-revolution or marching with the armies in the Vendée. Influenced by the demagogic rhetoric of figures such as Père Socle, *sans-culottes* burned priestly vestments on pyres as the National Guard camped in the nave of Notre Dame de Paris. Although the calendar had a series of antecedents, such as Maréchal's *Almanach des honnêtes gens* and the dating of the years of liberty, the hard work on its creation took place in the summer of 1793 against this backdrop of rising anticlerical feeling and increasing Jacobin confidence.

Despite this anticlerical context, the genesis of the republican calendar

6 N. Aston, *Religion and revolution in France, 1780–1804*, Washington, DC 2000, 210–11.

suggests that a concern for scientific truth and popular education provided its *raison d'être*, rather than an attack on religion, and its enforcement may also reflect a concern with centralisation as much as anticlericalism. Although Fabre D'Églantine was most popularly associated with the calendar, Gilbert Romme was in reality the primary figure in its creation.[7] While little is known of Romme's religious views at this point, he appears to have been deistic, rather than atheistic, in outlook and, as tutor to Count Stroganoff before the Revolution, attempted to provide him with an education in what would now be called comparative religion among his other studies. There is little in Romme's published speeches to place him among the nascent dechristianising or even anticlerical voices of the day. Politically, he was not associated with the populist Hébertist movement in Paris that would help to stoke the fires of dechristianisation. Moreover, his concerns while *en mission* were with Federalist revolts, rather than overtly Catholic counter-revolution.

Romme's thinking appears to have been driven by a number of factors, including a desire to excise the errors of the Catholic Church from the public domain, but this was by no means his dominant aim. He may well have wished to replace or destroy the hold of the Catholic Church on the popular mind, but, as Friguglietti argues, there is no direct evidence that anticlericalism formed his 'primary motive'.[8] Romme's *Rapport sur l'ère de la République* (20 September 1793) began by placing reform of the calendar within the context of the broader project for the systematisation of all weights and measures, a scheme 'of the greatest importance to the progress of the arts and of the human spirit'. Logically, progress demanded 'new measures of duration' freed from the errors that credulity and 'a superstitious routine' had passed down through the centuries. He asked whether a free people would want their history to be written on the same calendar as the crimes of kings and religious deceits. 'No', answered Romme, 'the Common Era was the era of cruelty, of lies, of perfidy, and of slavery. It ended along with the monarchy, the source of all our woes.' The report provided a catalogue of the errors of previous calendars, beginning with the Egyptians and the Babylonians, and summarised the history of Julian and Gregorian calendar reforms in a number of countries. Today, Romme suggested, one sensed the uselessness of these inferior reforms, to the despair 'of chronologists, of historians and of astronomers'. What was needed, Romme made clear, was a fixed annual point in the movements of the celestial bodies that could easily align the civil and solar year, without accumulating small errors over time. The

7 Friguglietti, 'Gilbert Romme'.
8 Ibid. 19.

solution was the autumnal equinox, happily matching the date of the inauguration of the republic. Under the new system, almanacs would no longer be packed with confusing dominical and golden letters. The old calendar had been a powerful tool for describing the world, one that priests had always successfully controlled and 'used to spread lies'. The new calendar, in contrast, would usefully spread truth, justice, and would encourage love of *la patrie* and all that led to prosperity.[9]

Romme's thinking seems to have been guided by several assumptions. Firstly, the need for a scientific calendar, free from the flaws and errors of the existing one. Secondly, he was aware of how the calendar had changed over time. Most of all, he rejected the historical, and arbitrary, links between the crown, Church and calendar. Romme's report promoted the new calendar as something scientific and 'pure', in contrast to the superstitions and irrationalities embodied in the Gregorian Calendar. Such rhetoric contrasted the benefits of Enlightenment, education and rationality with the corruption and obfuscation of the previous regime and mental order. The report associated the Church with the evils of corruption of the monarchy, but it found the calendar's greatest error to be its scientific failings. The Church was attacked within the report because of its support for the monarchy, its bizarre system of saints days, and for its incorrect and complex mapping of the religious and solar year. On balance, although destruction of religion was not Romme's primary aim, his project was certainly not hostile to the prospect. The tone of the debate in the Convention also points to growing anti-Catholic feeling: Pierre-Joseph Duhem, for example, worried that the proposed moral names could be corrupted, as had the calendar of saints. Intriguingly, he also cautioned that oath-taking priests wished to 'religionise our revolution'.[10] Overall, Romme's project is concerned with creating a calendar suitable for the republic, one informed by the latest and most accurate scientific reasoning. The Catholic Church and its calendar were attacked, but this was as much a function of the close links between the Church and what was seen as a flawed calendar and the general rhetoric of the day, which sought to explain in historical terms how the present had escaped the slavery of the past, as it was a guiding objective. This new map of time was guided by appeals to just science and a new nation, proposing a new age in which all religions were tolerated, but subject to civil control. Romme's later, related, project for a farmers' almanac showed the importance that he placed on practical reform, indebted more to Benjamin Franklin's *Bonhomme Richard* than

9 *Procès-verbaux*, i. 440–4.

10 Pierre-Joseph Duhem (1758–1807, member of the Committee of General Defence), 'Discussion sur le nouveau calendrier', *Le Moniteur*, 5 Oct 1793, 56.

to Maréchal's *Almanach des honnêtes gens*. Romme's report should not be read as a dechristianiser's charter or as a manifesto for cults of Reason or the Supreme Being; rather it was a tract authored by a *philosophe* in the age of Revolution.

Fabre d'Églantine's report ratcheted up the tension. Far more explicit in its attitude towards the Catholicism of the Gregorian Calendar, it made use of Fabre's rhetorical powers to concoct a charge-sheet against the Church. In its preamble, he quickly identified the Gregorian Calendar as 'the source of these religious errors'.[11] For Fabre d'Églantine, the old calendar was a commemorative instrument dedicated to saints, church rituals and the Nativity. The correction of errors now appeared to be the main focus of the reforms, compared to Romme's broader, more scientific project. The report continues with a long list showing how Catholicism had corrupted the calendar and undermined the virtue of the people. Through their encouragement of such a large number of ceremonies and feast-days, priests, these 'cruel enemies of human feeling and the sweetest of sentiments', contributed to moral corruption. Furthermore, they wished to subjugate 'the mass of farmers' by distracting them from the contemplation of nature and the correct observation of the seasons. The new names for each day of the year were 'more precious in the eyes of reason than some skeletons found in the catacombs of Rome'.[12] The language of the cult of Reason, and even of what became the cult of the Supreme Being, suffuses Fabre d'Églantine's text.

This change in tone and language from rationality towards moral sentiment reflected Fabre d'Églantine's intellectual style, but his disdain for the clergy also reflected a shift in the political mood. Widespread anticlericalism had developed into a positive programme of dechristianisation.[13] Revolutionary government had been declared. Two weeks after the publication of Fabre's report, Gobel's resignation as archbishop of Paris marked the beginning of a wave of dechristianisation. By removing saints' days and the familiar sequence of Christian festivals, the new calendar separated French men, women and children from the traditional way of ordering the year by liturgical events. Whatever its initial intent, the official disappearance of Sundays formed the most obvious breach with the routine of everyday life and the

11 Fabre d'Églantine, *Rapport*.

12 This is quoted in Schama, *Citizens* at p. 774.

13 Norman Hampson depicts the adoption of the calendar as the inflation of 'widespread anti-clericalism into a positive campaign for the dechristianisation of France' and 'a big step towards the elimination of Christianity from everyday life': *A social history of the French Revolution*, London 1963, 200–1. Hampson gives some support to the theory that the campaign was a political move by Hébert and his supporters (p. 202).

public face of the Church. As the Jacobin Joseph Chalier told the Convention, 'Through this new calendar, you have shown how much you want to kill fanaticism.'[14]

Dechristianisation

Despite Chalier's welcome, the evidence from the calendar's first few months is rather underwhelming in terms of its contribution to the dechristianisation effort. From 1792, and certainly by the autumn of 1793, Jacobins eagerly contrasted the reason and liberty embodied by the republic with religious 'fanaticism' and 'superstition', particularly that of the Catholic Church. Priests were suspected of supporting *émigrés* or of spreading counter-revolutionary sympathies, and were known to be involved in the rebellions in the Vendée. Revolutionary armies marched out of Paris with the intention of destroying the influence of the Church. This uncoordinated mix of anti-clericalism and popular violence has subsequently become known as 'dechristianisation'. Unknown in the 1790s, the term has come to be used by historians to describe a series of processes and events, beginning with attacks on the loci of ecclesiastical control and symbols of its power, such as tithe barns and rich monasteries, and culminating in the erection of Temples of Reason and the celebration of the cult of the Supreme Being in 1794. Largely the result of the demands of a Jacobin elite, it also drew on the iconoclastic fury of the Revolutionary armies and popular societies. Historians remain divided on the question of whether this attempt to destroy the cultural and moral influence of the Church and replace it with a new form of civic festivity, ritual and system of public virtue was politically orchestrated, or a popular response to the ideas of the Revolution, although many would agree with Vovelle's depiction of dechristianisation as neither quite an 'extraneous Machiavellian plot, nor the spontaneous flowering of a popular movement'. It was something altogether more complex, combining the actions of local representatives-on-mission, popular societies and revolutionary armies, all operating against the backdrop of a battle between Parisian political factions.[15] While they may agree on the general outline of what dechristianisation means, historians might be less inclined to agree with Vovelle's conclusions that the cultural revolutions of 1792–4 can be directly related to earlier patterns of secularisation during the eighteenth century and

14 This is quoted in Aston, *Religion and revolution in France*, 264.

15 Vovelle, *Réligion et révolution*, and *The Revolution against the Church: from reason to the supreme being*, Columbus, OH 1996, 151–3. The quotation is from *The Revolution against the Church* at p. 176.

that revolutionary dechristianisation left an indelible impression on the subsequent map of French religious observance. We remain uncertain about patterns of eighteenth-century belief and unbelief, and lack a full picture of the social meaning of religious ritual. Similarly, full-blown dechristianisation is all too easily ascribed to what Aston terms 'heightened expressions of anticlericalism'.[16] The conflicts over the observance of the *décadi* soon fell into patterns that would have been familiar to post-Tridentine religious and secular authorities.

Despite the easy links between the calendar's symbolism and republican religion, its dechristianising effects were relatively limited, at least if its decadary festivals are compared with the more violent destruction of churches or the arrest of priests. As Vovelle notes, the festival of the tenth day only occupied a 'modest place' within the addresses made to the national government by local groups. At best, it earns an 'arbitrary mention', such as this laconic note from Rouen (24 *floréal*): 'the decades were celebrated with all the pomp of the republic'. Elsewhere, communes reported just that 'the decrees and laws were read out on each *décadi*'. Dechristianisation, of course, was a complex, and remains a deeply ambiguous, phenomenon.[17] It arose from several quarters. The engines of popular revolution (representatives-on-mission, revolutionary armies and popular societies), together with *ad hoc* local gangs, were in the main responsible, with the intensity of activity varying from region to region. Visible, or aural, signs of religion and Christian piety, such as crosses or bells, were attacked or destroyed. Masses and other religious services were obvious signs of Catholicism and could be condemned. In the west the local population was informed that superstition, which fanned the flames of counter-revolution in the Vendée, still held sway over many communes and needed to be snuffed out. Similarly, observance of religious feasts and Sundays led to idleness. All good citizens had a duty to work, regardless of the old calendar.

The *décadi* could be used to attempt to stamp out the old religion. In Nancy, for example, on the 'second décadi of the Year 2':

> Confessionals were burnt at the foot of the scaffold of the guillotine, to universal acclaim from a procession of more than four thousand citizens, showing that the people of Nancy now are the most open to Reason and most energetic that the republic can offer. This festival ended with the burning of the letters of holy orders of a married priest [...] All day long, the air resounded with the repeated cries of *vivre la république*![18]

16 Aston, *Religion and revolution in France*, 260.
17 Vovelle, *The Revolution against the Church*, 176.
18 *Courrier républicain*, 30 *brumaire* II (20 Nov. 1793), 163.

As well as attempting to plant a new cult of Reason, officials recognised quickly that the use of the old calendar, particularly Sunday observance, could reveal an adherence to the old order of things. Jacobins might consider it important to stamp out such behaviour, or report it to the local surveillance committee (although, tellingly, there is little evidence of such activity). Departments were also asked to organise festivals of the tenth day and report on their success. Observation of 'formerly-called Sundays' and *décadis* did not pass unobserved by the authorities: the reports of *surveillance administrative* in series F^7 of the Archives nationales provide some information on the situation in a number of departments and municipalities. The revolutionary commission in the Somme reported with a certain bravado that subsequent to the decree of 5 October all the feasts and Sundays of the Catholic Church 'no longer existed' and that priests who 'preserved such prejudices' would be arrested, taken to prison and punished as law-breakers. The department forbade all religious feasts and ceremonies and publicly posted the proclamation for all to see.[19] But the words of a Jacobin *agent national* in the Gironde echoed a more familiar pattern: there were many in the Year II 'who do not solemnise the day of the *décadi* and only celebrate the day of Sunday'.[20] Many reported that the *décadi* was frequently ignored and that the citizenry continued to observe Sundays out of habit and ignorance of the reform. Several correspondents informed the authorities that working according to the *décade* was too exhausting, since rest days came only every ten days instead of every seven. The public prosecutor of the *Tribunal criminel* for the department of the Ain denounced one Valentin Dupont of St Martin: Dupont had publicly attacked the length of the *décade* because 'man and beasts cannot work nine days consecutively'.[21] In Havre Libre (the Revolutionary name for Roye in the Somme), the majority of the village who celebrated Sundays and other religious festivals with dances and games insulted the remaining 'patriots' who only recognised *fêtes nationales*. The patriots demanded external assistance.[22]

In contrast, in areas where popular societies had more influence or the local authorities made special efforts to enforce republican moral codes, those who observed Sundays and displayed a lack of respect for the *décadi* risked

19 *Fêtes décadaires, arrêté sur la calendrier républicain*, 5 *brumaire* II (26 Oct. 1793), AD Somme, L 421; 'Extrait du registre … de la commission revolutionnaire du départment de la Somme', 6 *brumaire* II (27 Oct. 1793), ibid.

20 Laporteries, *Agent national* of the commune of Martien, Gironde, to the comité de salut public, 23 *ventôse* II (13 Mar. 1794), AN, $F1^c$ III Gironde 8.

21 *Surveillance administrative*, 6 *messidor* II (24 June 1794), AN, $F^7$382.

22 12 *messidor* II (30 June 1794), ibid.

falling foul of the authorities. At Arras, the site of a giant festival celebrating the new calendar on 20 *vendémiaire*, servants taking any day but the *décadi* were liable to be arrested, while harvesters at Plaisance in the Toulouse region were arrested for stopping work on a Sunday.[23] It was more dangerous if one was already politically suspect, as one family near Bordeaux discovered. They were denounced to the *comité de surveillance* by the popular society at Trequier for 'walking insolently in the street' on the *décadi*, rather than attending the *fête décadaire*: they were also under suspicion for being related to *émigrés*.[24] The Jacobin authorities tended to interpret traditional patterns of work and rest as signs of religious fanaticism, as in the Donjon, for example, where the peasantry continued to 'follow Sundays in the old way'. The local *comité révolutionnaire* wished to draw attention to this 'new mania for fanaticism'.[25] In some communes, as in Montpezat in the Lot-et-Garonne, *cabarets* remained closed out of respect for the *ci-devant dimanche*, again arousing suspicion from local bureaucrats and the *société populaire*.[26] Such behaviour, a contemporary English observer concluded, 'would have been a certain mean[s] of drawing down the resentment of the predominant faction'.[27] In the Years II and III, despite violent and sometimes systematic efforts, the religious aspirations of the new calendar quickly faded into the fictions of administrative dictate where they did not fall under the greater imposition of the Terror. The new calendar could be used alongside existing dechristianising impulses to stamp out liturgical routines, but the creation of a vigorous, self-perpetuating and popular system of alternative festivals was another matter.

Alongside these outbursts of popular dechristianising activity, the calendar's *fêtes décadaires* provided a structure for the more mundane matters of national bureaucracy. From henceforth an individual's encounters with the state would be guided according to the new system, rather than by the Church as had tended to be the case in the past. Births, marriages and deaths would be recorded on the tenth day, as would the new divorces. Official announcements, including the *bulletin des lois*, which aimed to share the law code with the people, would also be read aloud. Local pride and practical issues raised their heads at this point. It was generally considered too onerous

23 Friguglietti, 'Gilbert Romme', 20.

24 *Surveillance administrative*, 6 *floréal* VI (25 Apr. 1798), AN, F^{7}382. The reports are brief and do not reveal exactly what course of action the local authorities took.

25 *Comité révolutionnaire* of the district of Val-libre, Tarn, 11 *thermidor* II (29 July 1794), AN, F^{7}3822.

26 *Comité de surveillance* of Montpezat, Lot-et-Garonne, 30 *prairial* II (18 June 1794), ibid.

27 Tench, *Letters*, 207.

a requirement for a ceremony to take place in every commune and so the government announced that *fête décadaires* need only be held in the chef-lieu of each canton. This was not the ideal solution, since the commune's citizens were required to travel a considerable distance in order to reach the festival. Such disparities could cause communal conflict. As Moreau, a deputy from the Yonne, explained,

> by demanding that fêtes décadaires are celebrated in the chefs-lieux of each canton, the resolution contradicts the equality of the rights of the citizens: it attacks political equality; all the advantages are for the chef lieu; and all the disregard for the small communes, who find themselves forced to make tiresome and expensive journeys: this measure is likely to increase jealousy.[28]

The dechristianisation process was not entirely destructive or purely violent, but was also an attempt to institute a new form of civic ceremony. Charitable donations, for example, were sometimes collected on the *décadi*. Dumont informed the committee of public safety on 24 *pluviôse* II (12 February 1794) that 'in the north east of France, I am establishing festivals everywhere that profit the poor on the days of the décadi. The one that took place in Abbeville brought in 900 francs. These festivals are necessary to erase the memory of what was known as Sunday'.[29] Romme's final letters to his young wife during his imprisonment in Paris reminded her to give alms to the poor every *décadi*.[30] In some areas of life at least – large urban festivals, Jacobin philanthropy, education and local administration – it seems that a new form of quasi-popular revolutionary religion was fostered as a direct replacement for Catholicism's social functions (if religion is the correct term). After Thermidor, one might have assumed that it would wither once dechristianisation had been left untended and the cult of the Supreme Being uprooted, but after a period of fallow, it again flowered with a somewhat bitter harvest.

Fêtes décadaires

Some three years after Robespierre's fall, after the coup of 18 *fructidor* V (4 September 1797), in which 177 deputies were removed and forty-nine election results were annulled, France had again a radical government, with little aversion to employing Terroristic tactics in order to achieve their political, neo-Jacobin, aims. A number of moves swiftly consolidated their position.

28 A. Moreau de l'Yonne, *Opinion sur la résolution relative aux fêtes décadaires*, Paris 1798, 5.
29 André Dumont, Amiens, to the comité de salut public, 24 *pluviôse* II (12 Feb. 1794), AN, F[1c] I 84.
30 *Les Martyrs de prairial: textes et documents inédits*, ed. F. Brunel and S. Goujon, Geneva 1992, 295.

Press laws of 19–22 *fructidor* (5–8 September) castrated right-wing publications, while military tribunals were commissioned to try to execute *émigrés* and conspirators. Troublesome priests could again be deported, and the Directory granted itself emergency powers to dismiss and replace judicial and other government officials. Political clubs were reopened for the first time in almost two years. The replacement of the royalist Franeois Barthélemy (who was deported to Guiana) by Philippe-Antoine Merlin de Douai and of Lazare Carnot by Franeois de Neufchâteau on 22 *fructidor* V (8 September 1797) confirmed the regime's radical stance. Merlin became minister of justice and subsequently minister of police, and Neufchâteau was appointed to the ministry of the interior. Both men were anticlerical supporters of the new calendar. Merlin held increasing power during the Year VI, and his hand in particular could be felt in the law of 14 *germinal* VI (3 April 1798), which regulated the strict observance of the calendar. Although Marie-Joseph-Blaise de Chénier had first proposed a system of civic festivals in every canton on 1 *nivôse* III (21 December 1794), they were actually instituted by Neufchâteau, reflecting his enthusiasm for antiquity and 'pagan pomp'.[31]

From the Year V (1797), France's military successes allowed the regime to focus on social and religious policy. Following the Treaty of Campo-Formio of 27 *vendémiaire* (18 October), France's strategic position was less precarious, and the Directory attempted to consolidate the radical gains of the Revolution with strict social legislation, including a raft of religious and calendrical laws. Religion was required to be subservient to the republic, and the calendar was one means of achieving this realignment of secular and ecclesiastical power. A new oath of allegiance declaring a hatred of royalty was instituted (some 10,000 priests refused to swear, making them liable to deportation) and church property was sold at cut-price rates in order to prevent its religious use.[32] Although there were striking local variations, the Directory's anticlericalism and anti-Catholicism continued to inform legislation, particularly as it related to the observance of the calendar. Soon after the coup of the 18 *fructidor*, Joseph Fouché disseminated a circular requiring the clergy to 'sanctify the *décadi*'. On 14 *germinal* VI (3 April 1798), the Directory decreed that all official acts had to be dated according to the republican calendar, arguing that observance of the calendar would help to bring about social order. The law impinged on a range of everyday activities. The decree began with a defence of the republican calendar and continued to

[31] Neufchâteau had also previously proposed civil calendar reform: *Mémoires de Barras: membre du Directoire*, ed. G. Drury, Paris 1895, 435–7.
[32] Doyle, *Oxford history*, 334.

prescribe its adoption by all public agencies, including *caisses publiques*, the postal and message services and public schools. Bourses, fairs, markets, contracts and agreements were to follow the *décade*, the new months and the *jours complémentaires* (the new title for the *sans-culottides*). The law now required the executive council of each department and municipality to take all measures 'to accelerate the changes demanded by the new division of the year'. All good citizens were to provide an example by following the new calendar in all their correspondence, whether public or private. The constitution also recognised only one calendar, in order to remove the 'last traces of the royal, noble and sacerdotal regime'. The decree denounced all those who regulated their meetings by Sundays and feasts of the previous calendar and ordered all public officials to regularise their audiences by *décade*. State schools were obliged to observe the *décadi*. It also decreed that canals, rivers and public carriages of all types were not to be used on the *décade*. Finally, dates on periodicals were to follow the new style, as the old era 'no longer exists for French citizens'.[33]

Yet, some suggested, even these requirements fell short of what was necessary for the enforcement of the republican calendar, and three further pieces of legislation were introduced. The law of the 17 *thermidor* VI (4 August 1798) required that shops and boutiques close their doors on *décadis* and national festivals. These restrictions were followed by the law of 13 *fructidor* VI (30 August 1798), which stipulated a *fête* for every *décadi* and enforced the *décadi* as the only day on which marriages could be celebrated. The law of 23 *fructidor* (9 September) confirmed the Directory's desire to forbid the old calendar 'in all acts or agreements, whether public, private, [in] periodicals, posters and writings'. The offence carried a fine of ten francs for ordinary citizens and fifty francs for government employees and notaries. Communes had to ensure that market days were in accordance with the rhythms of the new calendar.[34] The later legislation also demanded that teachers take their charges to the *fêtes décadaires*, and it stipulated that the school could face closure if it neglected these duties. Numerous circulars authorised by leading figures such as Fouché, Neufchâteau and Lambrecht were sent to departmental authorities, emphasising the importance that the Directory placed on the strengthened policy.[35] In one circular from the minister of police, it was even argued that the calendar laws were the foundations on which all other

33 *Arrêté du Directoire Exécutif qui prescrit des mesures pour la stricte exécution du calendrier républicain*, 14 *germinal* VI (3 Apr. 1798).

34 Ibid.

35 C. J. M., Lambrechts, circular, 29 *germinal* VI (18 Apr. 1798), *French Revolution Research Collection* (microfilm), Oxford 1990–8.

republican institutions could be erected, since the calendar informed peoples' opinions and morals.[36] Local help was also welcomed: the minister of the interior, recognising the difficulty of applying the law in all departments, actively sought advice from local officials on the best means of pursuing such a policy.[37] Such actions appear far more thorough than those of the Year II, and in part reflect the development of republican administrative structures as well as a more coherent religious policy.

The Directorials found the republican festivals in an enfeebled state in the summer of Year VI, perhaps in part due to the resumption in Catholic worship, largely spearheaded by the activities of female parishioners.[38] In the department of the Nord, for example, cantons were asked to report on the status of the *décadi*; the answers received were not particularly encouraging. One informed the *agent national* to the *directoire exécutif* at Douai that he could not 'leave him unaware that in general the *décadi* is only celebrated by a very small number of republicans [...] The rest continue to observe the *ci-devant* Sundays and several festivals of the Roman [catholic] calendar'. Religious sentiment was accompanied by 'traditional amusements' in the countryside on Sundays, 'in the old manner'. The author believed that enforcement of republican institutions was the only way to counteract such influences.[39] A handwritten codicil on a printed circular remarked that 'these old habits are difficult to vanquish'.[40] None the less, cantons closer to the towns appeared to have had some success, or at least felt it prudent to report the efforts made and the good behaviour of the citizens in their area. According to one report from the canton of Lannoy, to the north-east of Lille, an area long troubled by *émigré* priests, the *décadi* was 'religiously observed'. The streets were reportedly filled with people, wearing ribbons and their best clothes as they did on 'the greatest of the former festivals'. But, cantonal officials believed, success would only be assured if priests could be made to hold their services on the *décadi* and *quintidi* alone; only then would the old festivals and Sundays be forgotten in the countryside.[41]

Interest in dechristianising activity and the symbolism of national festivals has perhaps distracted historians from paying attention to the practical aspects of the republican takeover of religious activity. In particular, matri-

36 J.-P. Duval, circular, 26 *frimaire* VII (16 Dec. 1798), ibid.

37 Department of the Doubs to the municipality of Besaneon, 5 *floréal* VI (24 Apr. 1798), AD Doubs, L 380.

38 Desan, *Reclaiming the sacred*, 286–7; Aston, *Religion and revolution in France*.

39 *Observance des fêtes décadaires*, 22 *prairial* VI (10 June 1798), AD Nord, L 1265.

40 8 *floréal* VI (27 Apr. 1798), ibid.

41 15 *fructidor* VI (1 Sept. 1798), ibid.

mony became an entirely civil ceremony, but also one that was at the heart of communities, especially in the countryside. Along with births and deaths, which had to be publicly recorded at the *fêtes décadaires*, marriages could only be officially consecrated at a civic ceremony in accordance with the decision of the Council of the Five Hundred in the summer of Year VI. The government felt it important to provide all the principle ceremonies of one's life with an 'august and touching' quality.[42] Marriages, the politician Lemercier pointed out, were popular amongst the people, particularly in rural France, and ought to be used to support the republic whenever possible.[43] The republican calendar could be used to mark the most important stages of one's life, providing occasion for reflection on the past, the present and the future, and supporting the family, which was 'the image of the republic'.[44] Some worried whether weddings would undermine the solemnity of the *décadi*. In consequence, the betrothed were required in Dijon to be 'dressed decently, accompanied by their father and mother'.[45] There was a tension between the desire to engage with the community, which would have seen the wedding as the opportunity for celebration, and the solemn requirements of the ceremony. For example, one deputy, Jean-Marie Heurtault-Lamerville, thought that 'of the three great events in the life of man (birth, marriage, & death), marriage is certainly that which comprises and merits the greatest of solemnity'.[46] Despite these drawbacks, it was believed that 'the celebration of marriages in the *chief-lieu* of the canton can, at least sometimes, spread the *décadi* further'.[47] Such matrimonial strictures sometimes caused problems in more inaccessible areas, such as the Ile-de-Groix, off the coast near Lorient. Here, a special committee decided that it was difficult for citizens to be ferried across the channel to the canton of Port Liberté (the Revolutionary name for Port Louis) on the *décadi*. As a result, the islanders tended not to celebrate the Tenth Day, and this also meant that marriages could not officially take place,

42 *Rapport fait par Bonnaire du Cher, au nom des commissions d'instruction publique & des institutions républicaines, réunies sur les fêtes décadaires: séance du 19 messidor an 6*, Paris 1798 (AN, ADVIII 18 5).

43 L. N. Lemercier, *Opinion sur la résolution du 6 thermidor, relative aux fêtes décadaires*, Paris Year VI.

44 M. A. de Cornet, *Opinion sur la résolution relative aux fêtes décadaires*, Paris Year VI, 4.

45 J. B. Leclerc, *Rapport sur les institutions relatives à l'état civil des citoyens*, Paris Year VI; *Célébration des fêtes décadaires*, AD Côte-d'Or, L 476.

46 J. M. Heurtault-Lamerville, *Opinion de Heurtault-Lamerville, député du Cher & membre de la commission d'instruction publique, sur les fêtes décadaires: séance du 28 messidor an 6*, Paris 1798 (AN, AD VIII 18 15).

47 F. Guesdon, *Opinion sur le second projet de résolution présenté par Bonnaire, le 19 messidor, en ce qui concerne la célébration des mariages au chef-lieu de canton, les seuls jours de décadi*, Paris Year VI, 2.

since they were only valid if the ceremony was held on that day. It was decided that the *décadi* would be celebrated on the island as well as in the *chef-lieu* of the canton. In this way, concluded the Council of the Five Hundred which considered the case, neither the republican sanctity of the marriage nor of the *décadi* would be undermined.[48]

The Tenth Day could also function as an occasion for important communal business, as well as a demonstration of local or familial pride. The canton of Val Julieu used the opportunity of the *fêtes décadaires* to provide warnings of disease amongst the local cattle, six of which were infected at the time. In some areas the *décadi* was supported by the anti-royalist political societies often formed out of the rump of Jacobin clubs, known as the Cercle constitutionnel. In the Sarthe in 1798, for example, the Cercle in Le Mans travelled each *décadi* in procession to neighbouring towns or villages, where a liberty tree was planted and a new branch of the Cercle inaugurated.[49] Yet, despite the best – and persistent – efforts of the Revolutionaries, these offshoots of early republican optimism largely failed to be grafted onto the daily routines in much of France.

The reaction of the Church

The Church's response to these impositions ranged from acquiescence to outright hostility. Despite the regime's claims to religious toleration, the post-Fructidor Directory's increasing hostility drew condemnation from the Church. One anonymous pro-Catholic pamphlet, published in Anvers in 1799, considered the new laws concerning the *décadi* to belong to 'new pagans'. It reminded the clergy of the words of the church Father Tertullian: 'It is not permitted [...] to conform to the practices of infidels and to celebrate their festivals, be it for the love of pleasure, be it for worldly respect.' The faithful should avoid the new festivals and ceremonies. This pamphlet also demonstrates something of the pressure that Catholics were under to avoid dressing in their Sunday best and observing their religion openly, and it suggests that some Catholics were even following the advice of the 'new pagans'.[50]

The question of open Sunday worship also vexed the Constitutional Church. In the Year VII (1798), Claude Le Coz, the Constitutional bishop of Ille-et-Vilaine and afterwards archbishop of Besaneon, produced his *Observa-*

48 Minutes of the meeting of the Conseil des Cent-cents, 22 *ventôse* VIII (13 Mar. 1800), AN, C 561.

49 P. M. Jones, *The peasantry in the French Revolution*, Cambridge 1988, 245.

50 *Lettre de l'administrateur du diocèse de *** aux prêtres de ce diocèse*, Anvers 1799, 32.

tions sur la célébration du dimanche. In this work, the bishop argued that moving Sunday services to the *décadi* contravened the *Droits de l'homme* (article VII), the Constitution (article 354) and the law of 7 *vendémiaire* IV (29 September 1795). Furthermore, Sunday services and the *fête décadaire* were fundamentally distinct, one being religious and the other secular. 'O holy day!', Le Coz concluded 'O happy day! You will always be the true Sunday, the true day of the Lord, by our faithful observance.'[51] The publication of the tract caused his friends and Constitutional colleagues, such as Pierre-Phillipe Grappin, to fear for Le Coz's safety.[52] Despite this, Le Coz could not be dissuaded from printing and sending a copy to the Directory. In the end, he was simply sent before the local *juge de paix* who just asked him a few questions, before releasing him with an apology for the trouble he had been caused. Le Coz seemed more worried that he would offend Grégoire by attacking the *décadi* during a period when the Constitutional Church was trying to present itself as a true friend to the republic.[53] The *Annales de la religion*, Grégoire's journal and mouthpiece for the Constitutional Church, reveals something of this debate within the organisation. Several articles show that Sunday services were not only threatened by the government, but also by some clergymen who were tempted to move their services in order to be in line with the secular law. For example, in 1798 the bishop of Versailles urged priests to avoid antagonising republican officials and conflicts over the *décadi*.[54] In general, the journal that published Le Coz's article, and the Church, abhorred attempts to transfer or curtail their rights to religious freedom and the practice of a faith wholly compatible with the ideals of the republic. Grégoire's *Mémoires* condemned the transplantation of Sunday to the *décadi* as 'a tyranny' and as a stupid idea, since it failed to understand the spirit of the people and offered 'neither nourishment to the spirit, nor consolation to the heart'. The regime's methods, he noted, were at once 'shrewd and barbaric': shrewd, because poets 'had been invited to compose hymns for the ceremonies', which were combined with the 'magic of music'; barbaric, because priests were 'hunted, incarcerated and transported far overseas'.[55]

The response of the Church to the calendar during the Directorial regime varied throughout France, depending as it did on the complex relationship

51 A. Roussel, *Un Évêque assermenté, 1890–1802: Le Coz, évêque d'Ille-et-Vilaine, métropolitain du Nord-Ouest*, Paris 1899, 299.

52 Pierre-Phillip Grappin (1738–1833).

53 Roussel, *Un Évêque assermenté*, 300–3.

54 *Annales de la religion* (Year VI); Jean-Baptiste Flavigny, *Instructions pastorales sur l'accord de la religion catholique avec le gouvernement républicain, et sur la conciliation des devoirs essentiels du culte chrétien avec les lois de l'état en France*, Paris 1799, 58.

55 Grégoire, *Mémoires*, ii. 76–7.

between congregation and clergy, local politics and parishioners' piety. Efforts at a compromise between Church and State did take place, including attempts to encourage priests to move the mass from Sunday to the *décadi*. 'You must', proclaimed one such circular from the Doubs, 'urge all the ministers of religion to transfer their festivals and most important religious ceremonies to the *décadi*.'[56] In the department of the Nord, priests in the canton of Maroilles did indeed move their ceremonies to the *décadi*, most likely under pressure from local officials. Elsewhere, priests attempted to maintain their Sunday services.[57] The dual nature of the Church, with Constitutional and non-juring Catholic clergymen, further complicated the situation. (Constitutional bishops could not enter many 'White' areas for fear of their lives). Areas of counter-revolution or foreign interference tended to be zones of especial supervision. In the Nord, for example, the proximity of rebellion in the Belgian departments, where some 8,000 priests refused to swear the oath of hatred for royalty, ensured that the behaviour of priests, many of whom had been *émigrés*, was carefully monitored.[58] In the Year VI the department decided that 'it is time that we knew to what point we can account for the attachment of ministers of religion to the republican government'. It declared that in order for 'reason to triumph over ignorance' the entire clergy had to employ the republican calendar, and it demanded that they transfer their ceremonies to the *quintidi* or *décadi*.[59] Each commune had to report the response of the priests living in their area. Clerical reactions varied. Several priests replied that they had transferred their services or said mass everyday without a high mass on Sunday, but many others prevaricated or simply refused to transfer their offices. One of these, Francois Dupuis, a priest in Erquinghem-Lys, replied that 'nothing [...] is more essential to my religion than the sanctification of Sundays'. Like Le Coz, he maintained that the Declaration of the Rights of Man and the Constitution of the Year III supported his position, and subsequent laws also confirmed on him this 'inalienable right'. As a conciliatory gesture, he noted that he observed the *décadi* and other national festivals as well.[60] Other priests, such as two in Dourlers, believed that they did not have the necessary ecclesiastical authority to transfer Sunday services to the *décadi*. Five priests in Dunkirk asked for more time to reflect on the question.

56 Circular, Besancon, 17 *frimaire* VI (7 Dec. 1797), AD Doubs, L 380.
57 Department of the Nord to the comité d'instruction publique, 25 *prairial* VI (13 June 1798), AN, F^{17}1454.
58 Doyle, *Oxford history*, 334; *Observance des fêtes décadaires*, AD Nord, L 1265; *Fêtes républicain – décadi*, L 5046; *Cultes Belge*, L 5045.
59 Circular, 9 *fructidor* VI (26 Aug. 1798), AD Nord, L 5046.
60 *Procès-verbaux*, 27 *fructidor* VI (13 Sept. 1798), ibid.

Administrators might be exhorted to apply the Directory's religious policy, but not all thought that this made for prudent local politics. 'Far from me to comment, citizen, on the ridiculous idea of hampering the sacred right of all French citizens to the free exercise of the religion they choose', noted a *commissaire* in the Doubs. He conceded that, 'I know that no government can protect or prohibit a religious idea, whatever it may be.' Despite these reservations, the official continued to attack the canton of Mouthe for the ringing of bells on Sundays, which went against the law of 7 *vendémiaire* IV (29 September 1795).[61] More often, the record is silent, suggesting that the administrator in question thought it better not to inform Paris of the lack of enthusiasm for the new calendar and the persistence of religious observance. Perhaps perversely, the Directorials created a focus for counter-revolutionary activity. As the next chapter will show, this concern for a new regulation of time not only conflicted with faith, but with the requirements of labour. None the less, their policies continued under the Consulate and for a while were even pursued with greater vigour.

The Consulate

The religious question was perhaps the most urgent of the problems facing the Consulate, which came to power on the back of the coup of 18 *brumaire* (Year VIII/9 November 1799). Without some form of religious settlement, France would continue to be riven by continued social strife as White masses and refractory priests troubled local administrators. But between *brumaire* and *nivôse* of the Year VIII, far from attempting to abolish the *décadi*, the government attempted 'to perfect it'. Lucien Bonaparte, Albert Mathiez argues, resolutely defended the civic nature of funerals – and, by association, civil religion – and supported Fouché's attempts to control infringements of the law regarding behaviour on the *décadi*. The resolutely republican Mathiez possibly over-emphasised the importance of the *culte décadaire*, but he does provide evidence for continued government support for the *décade* and *décadi*. Lucien Bonaparte, for example, wrote to the *commissaire central* of the Yonne concerning some sixty to eighty workers and farm hands who were publicly working on the *décadi*. This, according to the circular, contributed to general disorder and undermined the *fêtes décadaires*. He invited the *commissaires* to take appropriate action to prevent this occurring again.[62] In many areas, repression of religious activity and enforcement of the republican calendar continued.[63]

61 Circular, 4 *frimaire* VII (24 Nov. 1798), AD Doubs, L 246.
62 Mathiez, *Théophilanthropie*, 597–600.
63 Sciout, *Histoire de la constitution civile*, 34.

However, evidence of police interference in decadal behaviour drops off completely after the Year VIII, despite the measures taken towards the end of the Directory. In Paris, *décadis* were well observed, at least in that people treated them as a day of leisure rather than another workday.[64] This may be seen as a trend throughout the Revolution, as *décadis* and Sundays both became occasions for sociability, particularly in Paris and other large towns. The social excesses of 'les gros', as the flamboyant rich were known, associated with the Directorial period, could flaunt their time as well as wealth: the *décadi* provided just such an opportunity.[65] Promenading was a popular activity for those with a certain amount of wealth. Dancing, drinking and game playing provided leisure opportunities for many, while all could afford a stroll and a gaze. In Toulouse, in the Year II, it was noted by the *Journal révolutionnaire* that although the churches had been closed, boutiques were ordered to be open and were frequented by women who, the writer complained, were neglecting their work. The streets and promenades were filled. The article did not wish to attack *liberté*, but argued that it was a time of Revolution and self-sacrifice.[66] Such stoicism was not shared by all. Journals and contemporary papers were soon filled with events that could be attended on the *décadi*, from balloon ascents to *bals*. Theatres were open on the *décadi*, although officially only plays, as officials in the Doubs warned, 'worthy of a republican people' could be performed, while the Ministry of the Interior was advised that a public guillotining ought not to go ahead because of the great number of people who promenaded on the *décadi*. While civic temples were deserted on the *décadi* except for wedding parties, the streets of Paris were bustling with people taking promenades, attending spectacles and, according to police reports, filling public spaces.[67]

Despite this, Sundays were given official sanction within months of Bonaparte's coup. On 7 *thermidor* VIII (26 July 1800) the *décadi* lost its status as the official day of rest for government workers, although marriages still had to be announced at the civic ceremony and fairs and markets continued to be arranged according to the republican calendar. Sundays regained their status as a day of religion, and Bonaparte admitted in 1804 that 1 January counted among 'les fêtes de famille' because of its popularity amongst the great majority of the French.[68] The customs of the Gregorian and Christian year

64 C. Brelot, C., *Besançon révolutionnaire*, Besançon 1966.

65 M. Lyons, *France under the Directory*, Cambridge 1975.

66 *Journal révolutionnaire de Toulouse ou le surveillant du Midi* liii, 2 ventôse II (20 Feb. 1794).

67 Circular, Besançon, 17 *frimaire* VI (7 Dec. 1797), AD Doubs, L 380; AN, AF III 47.

68 Circular, 18 Dec. 1804, *Correspondence de Napoleon Ier*, Paris 1858, no. 8222; Beck, *Histoire du dimanche*, 155.

had already begun slowly to reappear. In Besancon, in 1799, Christmas mangers were displayed and on 1 January 1800 confectioners were crowded for the purchase of New Year sweets. Yet, the return was not immediate. In 1802 the archbishop of Besancon noted that midnight mass had been celebrated in Paris in 1801, but so far this had not taken place in his city.[69] At the start of 1802 an English visitor in Paris was given 'a hundred whimsical little "bonbons" [...] as "Étrennes" or New Year's gifts'.[70] Sundays became the official day of rest for government workers on 18 *germinal* IX (8 April 1802). Over time, M. Dimanche had triumphed, perhaps as much as a day of leisure as a Catholic day of rest.

That the calendar's trajectory can often be linked to the political developments of the revolutionary and Napoleonic era serves to underscore the close connections between it and national symbolic language, but there are also a series of divergences and overlaps with other historical or local processes that deserve to be considered in more detail. As Noah Shusterman has shown, calendar reform, particularly the limiting and policing of religious festivals, was pursued with some vigour during the *ancien régime*, and religious holidays in particular attracted some considerable attention from the *philosophes*.[71] Official policy also took place within the context of local politics, which presents a complex pattern of accommodation, resistance and, at times, eager cooperation, and which were as much matters of policing as they were of pious or anticlerical sentiment. It is to this relationship between the timing of work, rest and recreation that we now turn.

69 Archbishop of Besancon to mayor of Besancon, 20 *frimaire* XI (11 Dec. 1802), AM Besancon.
70 Brelot, 'Besancon révolutionnaire', 201; Sadler, *Irish peer*, 31.
71 Shusterman, *Religion and the politics of time*, 8.

5

Work and Rest

The policing of work during the *ancien régime*

In the Year III, during a debate on the *décadi*, Joseph Terral reminded the National Convention of the moral dangers of the old festivals: 'Remember that under the *ancien régime*, the local *fêtes*, the days marked by dances, were generally the occasion for immorality, and the place of crime, especially in the Midi: fathers would keep their children away from them.'[1] It was not just parents who were concerned about the potential corrupting influence of such times.[2] Much police business had traditionally concerned itself with temporal regulations, and a real continuity in policy can be detected between pre-revolutionary and revolutionary policing of work.[3] In part this was a consequence of the departmental inheritance of the administrative and police responsibilities from the royal intendants and the *parlementaire* 'police' powers.[4] The *ancien régime* bequeathed to the republic a raft of regulations restricting Sunday trade, and the police enforcement of the *décadi* during the Revolution can be seen as the continuation of well-established methods of controlling work-time in the interests of social order and in order to support a particular moral viewpoint.[5] Drunkenness on the *décadi* affronted republican

1 J. Terral, *Réflexions sur les fêtes décadaires*, Paris 1795.

2 See also L. Febvre, 'Travail: évolution d'un mot et d'une idée', *Journal de psychologie normale et pathologique* xli (1948), 19–28; D. Roche, 'Le Cabaret parisien et les manières de vivre du peuple', in M. Garden and Y. Lequin (eds), *Habiter la ville Xve–XXe siècle: actes de la tables ronde organisée avec l'aide de la D.G.R.S.T. et de la mission de la recherche urbaine*, Lyon 1984, 234–5; D. Margairaz, *Foires et marchés dans la France préindustrielle*, Paris 1988; M. Vovelle, *Les Métamorphoses de la fête en Provence de 1750 à 1820*, Paris 1976.

3 *Délibération de l'administration municipale d'Amiens, pour l'observation uniforme du Calendrier Républicain*, 13 messidor VI (1 July 1798), 3–4, AD Somme, L 421.

4 I. Woloch, *The new regime: transformations of the French civic order, 1789–1820s*, New York 1994, 38.

5 Names of barmen serving drinks during the hours of the high mass, 1 Aug. 1700, and petition from pastry cooks demanding permission to open their shops from 9 a.m. until midday, 1 Mar. 1755, AM Dijon, D 29.

sensibilities, just as rowdiness outside the mass had affronted religious sensibilities during the *ancien régime*. Sometimes the comparison was explicit. In Amiens, in the Year VI, the municipal council cited the Estates General, which had decided that the 'wise Police rules' relating to public order should be kept. It listed a number of such regulations from the eighteenth century and explained that they now applied to the *décadi*, rather than Sundays. In such cases the old methods of regulation were simply transplanted to the new republic.[6]

In this way, republican policing can be placed in the context of the *longue durée* extension of state power described by de Tocqueville and others, which, from the late seventeenth century, showed a marked increase in time-related regulation.[7] In 1666 Colbert attempted to reduce the number of *fêtes chomées* as part of general reshaping of the French state. A modern state needed a sensible calendar as well as canals. Although nominally religious, such festivals were notorious in elite eyes for their promotion of popular dissipation and rioting, as well as encouraging the moral dangers of cabarets and 'lewd dances'.[8] As Philip Riley and others have shown, this development was in response to royal concern for public morality, but it was a reaction that took on its own administrative momentum.[9] Conduct books and sermons also point to the concern about such behaviour. The practical consequences of these administrative pressures can be seen in the Paris Châtelet judicial records, which reveal in detail the consequences of this ongoing campaign to control drinking or working on Sundays or saints days. The lists of offences and collections of police statutes record numerous defendants charged with a range of temporal offences, from late drinking, to selling produce during proscribed times or dates. From 1686 the stalls of Les Halles in Paris were forbidden from selling their wares on Sundays and feast days and their shop doors had to be fastened with a heavy chain. Other regulations were added

6 *Délibération de l'administration municipale d'Amiens, pour l'observation uniforme du Calendrier Républicain*, 13 *messidor* VI (1 July 1798), 3–4, AD Somme, L 421.

7 Williams, *Police of Paris*. See also M. Sonenscher, 'Work and wages in Paris in the eighteenth century', in M. Berg, P. Hudson and M. Sonenscher (eds), *Manufacture in town and country before the factory*, Cambridge 1983, 166; AN, Y 9498, 9499.

8 J. M. Ultée, 'The suppression of fêtes in France, 1666', *Catholic Historical Review* lvii (1976), 181–99. For an important example of the historical search for a transformation in economic systems in the shift from irregular rural festivals to the more regular, industrial patterns of Sunday observance see C. Hill, *Puritanism and revolution: studies in interpretation of the English Revolution of the 17th century*, London 1964, and Thompson, 'Time, work-discipline, and industrial capitalism', 56–97.

9 P. F. Riley, A *lust for virtue: Louis XIV's attack on sin in seventeenth-century France*, Westport 2001. For the legal context see H. Hauser, *Travailleurs et marchands dans l'ancienne France*, Paris 1929, 162–3.

over the years: in 1697 boutiques and linen shops also had to close on these days, and, from 1701, all Parisians were officially forbidden from working on Sundays.[10] Certain trades were allowed to operate on Sundays in Paris and its *faubourgs*. Butchers were authorised 'to open their stalls, to sell butchered meats […] on Sunday', presumably because of the importance of meat to the Sunday meal. They were to close on Saturday instead, possibly because of a demand for market space. Regulations did not only relate to dominical or feast-day observance, but were matters of public order. Citing a 1712 proclamation forbidding the sale of any merchandise on Sundays and holidays 'on bridges, quays, footpaths, and under all the gates', the Police du Châtelet were informed in 1732 that 'many continue with greater scandal than ever to set up stalls of all sorts'.[11] Artisans' hours were limited in Paris: work could not be undertaken by candlelight to avoid poor workmanship.[12] Nor were controls limited to the capital. In Dijon, for example, the majority of trades and merchants were forbidden from working or selling on Sunday. Public drinking and dancing was also controlled. It was forbidden for

> all restaurateurs, innkeepers and cabareteurs and all sellers of coffee and liquors, to provide drink and food in their abodes, houses, cabarets and cellars during the parishes' grand masses and vespers on Sundays and holidays, on pain of a 20 sols fine for any person found in their said houses, inns and cellars.

Bakers and pastry cooks were allowed to advertise and sell a certain amount of bread and cakes, but they 'must remove the notice of closure from their shops, and put in front, outside the times of the said services, two loaves of bread and several cakes just to make themselves known'.[13] Charges were brought against those who infringed the drinking regulations, although the restrictions were relaxed somewhat in the mid-eighteenth century, when baked goods and bread could be sold from 9 a.m. until midday (it seems that drinking restrictions remained). In all these instances, the law attempted to prevent people from making a living, at a time when few could afford the pleasures of leisure time; such workers may have welcomed one day of enforced rest in ten compared to the more common *ancien régime* feast-days.

Restrictions were not as strict in the surrounding countryside. Sundays afforded an alternative to work in the fields and the vineyards and the chance to come to the local town, where stalls and shops would be open. This regula-

10 Riley, *Lust for virtue*, 130.

11 30 May 1668, 14 Aug. 1731, AN, Y 9498.

12 M. Slavin, *The French Revolution in miniature: section Droits-de-l'Homme, 1789–1795*, Princeton, NJ 1984, 44.

13 Minutes of *chambre du conseil* of the city of Dijon, 31 Mar. 1700, AM Dijon, D 29.

tion caused complaint from the clothiers and haberdashers of Dijon, who were required to shut up shop on Sundays in 1755. It was suggested that the opening of the shops was the cause of troublesome country folk coming into town.[14] Drinking during the hours of mass and vespers also continued to cause men to be brought in front of the town authorities and fined; such infractions do not appear to exist for shopkeepers and merchants. Either the records have been lost and the laws were perfectly adhered to, or, as the clothiers and haberdashers suggest, commercial pressures meant that some opened, at least for part of the day.

Sundays and feast-days therefore present a variety of faces to the historian. Religious legislation ensured that the seventh day was seen as a day different from the rest of the week, but there are also hints that such legislation was circumvented in many cases; commerce and Church could coexist in practice. Furthermore, the extent of Sunday restrictions varied through the day: the hours between mass and vespers were defined as especially sacred and attracted the strictest sanctions, thereby allowing activities to take place outside these times. Clocks and bells, for example, regulated drinking times, restricting drinking on Sundays – but also permitting it at certain times.[15] As Mercier commented on habits outside of Paris, 'In a few provincial towns the bells are sounded at ten o'clock in the evening to warn the drunkards to leave the bars. After a quarter-of-an-hour, all friends of Bacchus that may still be in the tavern are thrown out by the patrol.'[16] Presumably, if the patrol did not arrive, Bacchus's companions could continue in peace. Before the Revolution, drink sellers were supposed to cease trading by 11 p.m., and the police occasionally had to explain to tardy publicans that the time was measured by the local bell or the official's watch, and not by the drinkers. The Châtelet police records show that similar measures were in place in Paris, but the detail of such regulations differed throughout France, as did much temporal legislation: for example, sales of drink were ordered to cease in Clermont Ferrand by 9 p.m.[17]

Although religious and civil legislation protected the sanctity of the seventh day, in reality parishioners were rarely compelled to attend either mass or confession, and Sundays began to have nascent civic and social connotations as a day of leisure or family for many from the mid- to late

14 Clothiers and haberdashers of Dijon to the mayor and aldermen of Dijon, 24 Jan. 1755, ibid.

15 Riley, *Lust for virtue*.

16 L.-S. Mercier, *Le Nouveau Paris*, Paris 1798–9, vi.190.

17 AN, Y 12503; P. Bourdin, *Des Lieux, des mots, les révolutionnaires: Le Puy-de-Dôme entre 1789 et 1799*, Clermont Ferrand 1995, 14.

eighteenth century. Increases in wealth, the rise in domestic sensibility, and even changes in town planning, which gave increased space for activities such as promenading, all served to change the seventh day's character.[18] Policing also reflected this change, as long as such behaviour was well-ordered and 'dignified'. Indeed, one manual suggested, with some liberality, that 'gaming is a remedy, a rest that it gives to the spirit'. It continued, '[whether] this recreation takes place at home with one's family, or some friends, at an early hour, there is nothing there which harms the purity of morals, nor anything which inconveniences public order'.[19] Similarly, the model of de Toquevillian centralisation needs to take into account the many local accommodations and time-honoured irregularities in the face of central control. The *Almanach nécessaire*, a Parisian almanac from 1789, reveals that many government officials had their office hours on Sundays. The same almanac also acted as a useful guide to the many social occasions that were regularly held on Sundays, such as spectacles organised at the 'Redoute Chinoise', or other theatrical performances. [20] Sunday afternoons provided the opportunity for a number of social and leisure activities, as well as routines such as washing. Dancing, drinking or promenading provided other options, relatively irrespective of income.[21] When a Yorkshire gentleman, Jonathan Gray, briefly ventured to France in May 1814, he noted 'that Sunday was never observed by the shutting of shops, even before the Revolution'.[22] It may be that for some the republican calendar upset patterns of leisure and sociability, rather than simply of religion. It also attempted to overthrow complex routines in a matter of months or years, rather than decades.

The motive for reform may also be found in the emerging discipline of economics. Many Enlightenment thinkers considered the existing system of markets to be an inefficient remnant of a medieval past that constricted efficient trade. The *Encyclopédie*'s entry on markets, written by Turgot, estimated that twenty-four days were lost in this manner at a cost of 96 million *livres* (although it noted, even-handedly, that some of these would fall on

18 Beck, *Histoire du dimanche*.

19 N. de La Mare, *Traité de la police, où l'on trouvera l'histoire de son établissement, les fonctions et les prérogatives de ses magistrats, toutes les loix et tous les réglemens qui la concernent: on y a joint une description historique et topographique de Paris … avec un recueil de tous les statuts et règlemens des six corps des marchands et de toutes les communautez des arts et métiers: par M. Delamare*, i, Paris 1729.

20 *Almanach nécessaire, ou porte-feuille de tous les jours, utile à toutes personnes*, Paris 1789.

21 Mercier, *Tableau de Paris*, iv. 161.

22 Jonathan Gray, Calais, to Mary Gray, Paris, 24 May 1814, in *Papers and diaries of a York family, 1764–1839*, ed. E. Gray, London 1927, 145–52.

Sundays). Arthur Young, the English agronomist, deplored the loss to agricultural production resulting from the French peasantry's fondness for attending a market one day a week. He considered that such activities wasted the time of the small producer, who would squander an entire day selling a dozen eggs or a chicken, the price of which would in no way compensate for the day's labour lost.[23] Feast-days were also viewed as potential problems. Mercier stated that in Catholic countries – including, of course, France – '*fêtes* occupy a quarter of the year'.[24] This was a view that informed many Revolutionaries' attitudes towards feast-days. To such existing economic theories, the revolutionary period also brought the urgency of war, and an increasing animosity towards religion. Marcox, for example, worried that a multitude of *fêtes* harmed agriculture.[25] In theory, the republican calendar offered an answer to this problem as it reduced the number of *jours de repos*. In practice, however, as others pointed out, the net result was an increase in days off, since both the old and the new rest days were observed. Citizen Saint-Aubin, a teacher of commerce, mathematics and foreign languages, gave the example of the 'cordonnier dimanchier' who had no trouble persuading 'son compagnon décadaire' to celebrate Saint Monday with him. The situation would be reversed on the *primidi*. One reckoning suggested that such lack of productivity posed an annual cost of 160 million *livres* to the nation.[26]

The concern with days lost to labour may have been as much a rhetorical or moral concern as an economic one, but there were also moves to make such rest, or leisure, productive for the citizen and nation alike. For thinkers, idleness was to be decried (unless perhaps one was exercising the arts of conversation and sociability).[27] Dignified contemplation and civility were to be encouraged. The *décadi* reduced the loss of industry, but also encouraged appropriate use of free time. Well-organised fêtes would, it was felt by their proponents, encourage good citizenship, rather than the drunkenness and disorder that feast-days and Sundays were infamous for; this can be seen as part of the Jacobin 'war on idleness'.[28] In this manner, *décadis* were intended

23 A. Young, *Travels in France during the years 1787, 1788, 1789*, Gloucester, MA 1976, ii. 757.

24 Mercier, *Le Tableau de Paris*, 293; J. Faiguet de Villeneuve, 'Fêtes des Chrétiens', *Encyclopédie*, vi. 565–73.

25 J.-B.-P. Marcoz, *Objet et ordre des fêtes décadaires de la république franeaise*, Paris Year III, 2.

26 Saint-Aubin, *Réflexions*, 6.

27 D. Gordon, *Citizens without sovereignty: equality and sociability in French thought, 1670–1789*, Princeton 1994, 129–76.

28 Mathiez, *La Vie chère*, 443.

to improve the quality of time: hence the numerous plans and schemes for efficacious uses of the day, beginning with public singing and speeches and ending in games or in more gentle relaxation.

For the *philosophe* and the theologian alike, labour was ascribed a moral value, even if systems of morality differed.[29] Rousseau believed that manual labour brought an individual as close as was possible to the natural state, and helped to counter the dangers of idleness: Emile was to have a trade, rather than being a poet, musician, actor or writer, since those occupations associated with creativity, like study, 'wear[s] out the human machine'. The Church, of course, had long developed what Jacques Le Goff has termed 'a positive theology of labour', in which it found spiritual virtue in work.[30] Scholars of the world of the late eighteenth-century artisan have also been keen to stress how morality and work were intertwined in the day-to-day reality of the workshop. Social status was bound up with the morality of hard, and skilful, work. Work – or idleness – also provided one of the more common complaints by masters about their indolent, and hence immoral, workers. In comparison, journeymen counterpoised, in Michael Sonenscher's words, 'the frequent absence of masters from their *boutiques* in pursuit of clients or credit or both. The morality of industriousness was the medium in which the struggle to which time was put found its expression'.[31] The culture of eighteenth-century work, such as that revealed in Ménétra's diary, formed an intricate pattern, a 'penumbra of customary practices' tensely balancing wages, demand, credit and capital with the erratic schedules of artisanal production and agricultural work.[32]

The policing of work during the Revolution

From the late winter of 1798, those who worked openly on the *décadi* risked a knock on the door. Across France, republican officials began to make arrangements to patrol their section or commune, in order to catch those failing to observe the new day of rest. Those whom they found working in public view risked a summons to the civil court. These included an Avignon silk-worker, Marie Arnoud, who in *ventôse* VII (February 1798) was fined five

29 A. Jacob, *Le Travail, reflet des cultures du sauvage indolent au travailleur productif*, Paris 1994.

30 Le Goff, *Time, work and culture*, 58–70. On the history of the theology of manual labour see E. Whitney, *Paradise restored: the mechanical arts from antiquity through the thirteenth century*, Philadelphia 1990.

31 Sonenscher, 'Work and wages', 106, 109.

32 Ibid.

centimes for working in her shop. That same month, the Avignon police fined Madeleine Papolas, Louise Merindol and Marie Ceylan 30 *sols* apiece for working publicly on the tenth day in contravention of the law of 17 *thermidor* VI. The campaign to enforce the calendar continued throughout the coming year. On 20 *fructidor* (6 September), Elisabeth Jullian was found to be working, and even though she claimed that her doors and windows were closed, she was brought before the section's *juge de paix*. During the same court session, Marie Escoffier, charged with the same offence, also admitted that her workers were washing silk, but argued in her defence that her shop was closed to customers. Another defendant, Jean Barracan, denied that he himself had been working, but did go so far as to admit that his workers had done a little work on the morning of 20 *fructidor*. Raymand Chabeze was also brought to the *juge*, where he pleaded that it was true that he was preparing 'merchandise to go the market the next day', but, he argued, he only worked in the evening and did not occupy the whole day. He claimed that he was too observant of the laws to have done so; the authorities were not sympathetic and fined him.[33] In *vendémiaire*, in the nearby canton of Apt, east of Avignon, the *commissaire de police* walked the streets to check that the *décadi* was being correctly observed. He found a number of citizens working in public view and charged them accordingly.[34]

What can we make of the geography of reform? The unfortunate silk-makers arrested in Avignon worked in a predominantly republican city, whose pro-Revolution officials were conscious of the area's historical association with the Catholic Church; the surrounding area was also rife with counter-revolutionary activity. As a result, the municipal authorities were perhaps especially alert to policing potential signs of counter-revolution. The police archives and the records of the *juges du paix* in Paris, Nantes, Besaneon, Lille and Amiens reveal similar administrative efforts. All these towns had particular reasons to enforce the Directorial legislation on the 'strict observance of the décadi', notably their proximity to areas of counter-revolution or, in Paris's case, because of relatively well-developed policing and the ever-present eyes of cohorts of section activists. Elsewhere in France there was often less impetus to enforce the regulations on decadary working. Such patterns are possible to discern from the *comptes décadaires* – detailed accounts of the previous ten days sent by departmental officials to Paris and collated by the central administration. The reports suggest that in the majority of departments, officials were aware of the need to regulate rest days,

33 *Justice de paix* of the canton of Avignon, AD Vaucluse, 18 L 24.
34 24 *fructidor* VII (10 Sept. 1799), ibid.

but that many were happy to report that 'good citizens' observed it, with a few trouble-makers still keeping the old days of rest.

The archives also, perhaps unsurprisingly, show that particular attention was paid to urban rather than rural areas. Such lacunae were largely a consequence of the lack of administrative manpower and the differences between patterns of work in town and country, but they also reflected a special, long-standing concern with public behaviour in towns, where population density (including a large transient population), threats of dearth, criminality, riot or protest posed potential problems for the authorities. As the number and content of administrative records concerned with local policing suggest, this balance between municipal authority and the urban population was a delicate one, needing constant calibration.[35]

During the Revolutionary period, many republican officials would surely have agreed with *citoyen agent* Balland's report from the Doubs that 'the majority of the inhabitants [...] occupy themselves with the cultivation of their vines and their fields, and none observe the *décadi*'.[36] Sundays rather than *décadis* remained the common day of rest in this part of France, in line with the strong Catholic sentiment in the region. A number of letters from local agents agreed that infractions against the observance of the *décadi* had taken place and ought to be stopped.[37] Some agents were slightly more optimistic, such as in Rougemont in the Doubs, where the citizens had begun 'slowly and with a lack of concern to observe the *décadi*'. The *commissaire* who filed this report put the blame firmly at the doors of the Church and, more specifically, on refractory priests who preached that Sundays ought to be the principal day of worship.[38] Schoolteachers were also failing in their task of escorting their charges to the weekly *fête*. In consequence, a special watch was arranged for the 30 *ventôse* VII (30 March 1799). It revealed that citizens were in general occupied with the cultivation of their vines and fields. Few paid any attention to the fact that it was the *décadi*, despite the local government agent's report that the tenth day had been declared the day of rest and recreation.[39] Many other reports also pointed out the difficulty posed to the calendar by the tenacious hold of Sundays. Such lack of concern for the

[35] Williams, *Police of Paris*, 202–28; Schneider, *Ceremonial city*; Garrioch, *Revolutionary Paris*, 115–41.

[36] Municipality of Rougement to department of the Doubs, 1 *germinal* VII (21 Mar. 1799), AD Doubs, L 380.

[37] Correspondence concerning public festivals, ibid.

[38] Canton of Rougemont to the department of the Doubs, 24 *brumaire* VII (14 Nov. 1798), ibid.

[39] Balland, canton of Rougement to the department of the Doubs, 1 *germinal* VII (21 Mar. 1799), ibid.

calendar extended to officials in the neighbouring department of the Côte-d'Or, where *juges de paix* were criticised by the justice minister for failing to convict people charged with working on the *décadi*.[40]

In contrast, the evidence from Nantes points to a willingness to work with the law, rather than ignoring or being unaware of it, as seemed to be the case in Avignon and Besancon. Traders applied for permits to work on *décadis*. Pierre Auguste Deuviez, a Nantais pottery manufacturer, appealed to the administration for exemption from observing official days of rest. It was not possible, Deuviez argued, to halt production on the *décadi* because of the nature of his work.[41] The administration granted permission, but only on the condition that the manufacture did not close on Sundays and the old feast days. In contrast, a mussel-seller from outside of Boussay made the same request, but revealed a use of the old calendar at the same time as the new. When the *décadi* fell on a Friday or Saturday, the mussels could not be sold until the following week, meaning the loss of all the stock. She asked if produce could be sold on the at least the mornings of those *décadis*. The administration was not at all sympathetic to this request and answered that she had nine days out of ten to sell. She had to follow the law.[42]

Although the law was not applied consistently, there is evidence that this resulted from a certain sensitivity to the case in question; perhaps pottery was more important to the nation than mussels. More critical was the evidence that the mussel vendor provided of the existence of the old hebdomadal and religious practices and lack of sympathy for the new calendar. A number of reasonable conjectures explaining differences in restrictions can be made. Permission for necessary work, such as the need to launch a ship on the high tide, which coincided with the 10 *messidor* VI (28 June 1798), was granted.[43] Similarly, various sorts of traders were granted permission to open on the *décadi*. A newsagent was granted permission to open his *cabinet littéraire* on the *décadi* and official festivals so that periodicals and subscriptions could be collected.[44] Trades considered essential such as bakers and grocers, even *perruquiers*, were allowed to operate on the *décadi*. Vegetable, fruit and milk markets could also take place.[45] In contrast, non-essential trades and activi-

40 Lambrecht, minister of Justice, to the department of the Côte-d'Or, 1 *messidor* VII (19 June 1799), AD Côte-d'Or.

41 Pierre Auguste Deuviez to the canton of Nantes [Year VI?], AM Nantes, I2 8 9.

42 Mussel-seller, Boussay, to the commune of Nantes, 11 *prairial* VI (30 May 1798), ibid.

43 René Moiret, Nantes, to the commune of Nantes, 8 *messidor* VI (26 June 1798), ibid.

44 Permission to open *cabinet littéraire*, 6ème *jour complémentaire* VII (22 Sept. 1799), ibid.

45 *Affiche* of meeting of the administration municipale de canton de Nantes, 14 *floréal* VI (3 May 1798), ibid.; P. Clémendot, *Le Départment de la Meurthe à l'époque du Directoire*, Raon-l'Etape 1966, 257.

ties deemed to pose moral dangers, such as *salles de danse*, met with short shrift and were ordered to close. In enforcing this aspect of the law the administration worked to distance itself from the *police ecclésiastique*, who, it was argued, closed dances and condemned gaming in order to protect the superstitious sacredness of Sundays. In contrast to their *ancien régime* forebears, the republican legislators claimed that they wished to reprimand the entrepreneurs and increase public morality and sobriety. This may have been splitting hairs. Officials were also wary of inhibiting free trade and commerce.[46] There is something of a contrast between the two regions, resulting from the different responses to the Revolution and varieties of social and economic structure. Nantes possessed a more active and socially prominent bourgeoisie, with different patterns of trade compared to the predominantly rural Franche-Comté; the experience of counter-revolution also made the authorities alert to signals of opposition to the regime.

Police watches specifically aimed at ensuring the observance of the new calendar were also arranged in Amiens, and it seems likely that similar efforts were made in towns across France.[47] Measures were taken in a variety of cantons and departments to detect and discourage infractions of the law. In the Meuse, for example, the canton of Clermont forbade 'grocers and chemists' from opening their shops on the *décadi*, and the rural police (*gardes champêtres*) in the canton of Verdun recorded their dismay at a double desecration: those who 'work the day of the *décadi* in cleaning (that is to say, washing) their cattle, in clearing their dung from their doors'. The administration in the canton of Jametz was even more assiduous and attempted to forbid private written use of the Gregorian Calendar.[48] In the Sarthe, squads of eight soldiers were placed on the roads coming into Sablé to arrest those attempting to work.[49] Paris also provides evidence for policing of the tenth day. In the Year VI a special watch was arranged to arrest those working on the *décadi*. It went about its business on at least seven occasions in the Pantheon section of Paris, issuing fines where it could.[50]

The new measure of time became a means of rooting out the outsider – those who rejected, or who did not wholly adhere to republican ideals. Men

46 Department of the Loire-Inférieure to the canton of Nantes, 3 *prairial* VI (22 May 1797), AM Nantes, I2 8 9.

47 Report by Gilbert, *commissaire de police*, 'pour maintenir le bon ordre et faire exécuter le nouveau calendrier républicain', 11 *thermidor* VI (29 July 1798), ibid.

48 Aimond, *Histoire religieuse*, 440 n. 6; 10 *germinal* VI (25 Mar. 1799), AD Meuse, L 1474; Aimond, *Histoire religieuse*, 440 n. 2.

49 M. Reinhard, *Le Département de la Sarthe sous le régime directorial*, Saint Brieuc 1936, 350.

50 *Procès-verbaux des commissaires de police*, Years IV–VII, APP, AA 202.

and women suspected of counter-revolutionary activities could be attacked for their behaviour on the *décadi* and their lack of enthusiasm for national festivals; religious observance could also bring condemnation or arrest. Complaints were not just made about suspicious citizens within the community, but also against other communes. Those that observed the *décadi* were undermined by their neighbours who could be lax, even counter-revolutionary, in such matters. Although intended as a national calendar, emphasising the unity of the nation in time, in its practical expression it often instead emphasised the differences between groups and communities in French society. Often, the calendar emphasised the traditional and deep-rooted divisions, grudges and feuds of rural France. For example, by the Year VII, urgent or necessary work could be officially sanctioned by the authorities, but this sometimes created jealousies between different classes. Ploughmen in the canton of Messigny in the Côte-d'Or were given permission to work, leading to grumbling amongst the wine growers of the region who were liable to fines for dressing their vines on the *décadi*.[51] Such complaints of unfair competition were also common in the Nord. In Gevrey, just south of Dijon, a number of measures were put in place to encourage observance of the *décadi*, such as the reading of the laws on the *décadi* announced by a drum roll, but these efforts were undermined by those who simply continued with their usual business and ignored the festival. The *société populaire* noted that 'there exists in our canton a lack of authority [*difformité de conduite*] that chokes the spirit of society'. They petitioned the departmental administration to 'abolish this kind of schism' on the question of whether one 'occupies oneself with work on the day of the *décadi*'.[52] The canton continued to make complaints about those who lacked enthusiasm in their observance of the work restrictions on the *décadi* and hence set a bad example for the more easily influenced citizens. In *brumaire* Year VII (November 1798), Gevrey complained about non-observance of the *décadi* outside of Dijon, where a lack of national guardsman undermined administrative efforts to enforce observance. The result, argued the canton, was that the good citizens of Gevrey were being put at a disadvantage.[53]

Unfair economic advantage for one area (by working when others did not) and general anti-republican behaviour provided two, possibly complemen-

51 Canton of Messigny to the department of the Côte-d'Or, 2 *fructidor* VII (19 Aug. 1798), AD Côte-d'Or, L 476.

52 *Société populaire* of the canton of Gevrey to the department of the Côte-d'Or, 20 *frimaire* II (10 Dec. 1792), ibid. L 1660.

53 Canton of Dijon Extra-Muros, 25 *floréal* VI (14 May 1798), and canton of Gevrey to the department of the Côte-d'Or, 21 *brumaire* VII (11 Nov. 1798), ibid. L 476.

tary, motives for complaint and outrage on behalf of the local popular society or administration. Class tensions and roles were also apparent, at least in some parts of France. In Haubourdin, in the Nord, the *tribunal de police* tried ninety-two cases of those who were working on the *décadi* in the autumn of the Year VII, charging malfeasants three days of labour. But, in the majority of cases, the *fabricants d'huiles* complained that their workers refused to stop their labours on these days, since nearby cantons were also working. One citizen, who owned two mills, told his workers to leave, and that he, not they, owned the mill. The workers threw their keys on the floor and threatened to leave their jobs for good. The authorities took the view that the owner was responsible for his workers, as he was for his wife, his children and servants.[54] Government officials could be singled out by the community for attack: municipal agents in the Nord had to endure verbal assaults as they attempted to enforce the *décadi*, and one official had his house set on fire.[55] The calendar thus had the power to heighten social distinctions, whether those of politics, geography or trade. Such local conflicts were especially intense when it came to markets, the key points of sociability and trade across France.

Markets

The Directory saw the regulation of markets as a key component in the enforcement of the calendar. In *vendémiaire*, Year II, a decree required each department to draw up a table of fairs and markets according to the new calendar, with mixed results. Some towns and departments moved quickly to implement these reforms, such as Amiens which moved its horse market from Saturday to the *décadi* in *nivôse*, Year II (January 1794). Yet, as several almanacs record, many towns retained their old market days even after the new calendar had been introduced. Here again, the calendar led to local disputes as neighbouring communities often found themselves in competition with each other as they attempted to hold more popular or frequent markets, as well as restrict or allow labour.[56] The Directory attempted to stamp out these old inconsistencies. Article V of the law of 23 *fructidor* VI ordered all departments to redraw their market timetables in accordance with what was now termed the *Annuaire de la République* and forbade the staging of markets and fairs on the *décadi* and on days of former festivals.[57] The interior and justice

54 *Observation des fêtes décadaires*, 22 *vendémiaire* VII (12 Oct. 1798), AD Nord, L 1265.

55 10 *ventôse* 6 (28 Feb. 1798), ibid. L 1267.

56 Correspondence concerning fairs and markets: AD Somme, *Extrait du registre aux délibérations de la commune d'Amiens*, 21 *nivôse* II (10 Jan. 1794), AD Doubs, L 316.

57 *L'Ancien Moniteur*, xix, Paris 1863, 150–1; *Loi contenant des dispositions nouvelles pour*

ministers underlined the importance of the new measure with a series of circulars and set about collecting data from each department to ensure that they were complying with the new, secular scheme. Initially, the interior minister was saddened to discover that 'almost everywhere', fairs and markets followed the old calendar, 'each day of which suggests the idea of a dominant religion, whereas the republic recognises none; its principles alone must predominate'. He urged officials to counteract 'the perpetuation and growing restlessness of fanaticism', and to 'fix fairs and market days according to the republican era', believing that this measure would efface the 'final traces of these peculiar customs' of the old calendar.[58]

Romme and Fabre d'Eglantine had also stressed the burdens that the existing calendar placed on commerce and industry: the final report on the calendar contrasted the superstitious tricks of the Gregorian Calendar, which led to ignorance, with the supposed intellectual rigour of the new system, drawing attention to the benefits it would bring to the rural economy.[59] French agronomists under the Directory also attacked the superfluity of fairs and markets and the increase of holidays generated by the two calendars – the Gregorian and the republican.[60] Jean Baptiste Seraphin Villèle, the agronomist and politician, despaired at their number: 'too many fairs and markets will be especially harmful to agriculture by too frequently disrupting the work in the fields of farmers who must avoid [...] the loss of their time'.[61]

Here again the Directory intervened in a system of exchange that was not simply economic, but which formed a crucial part of the cultural and social fabric of French life. The pattern of work and rest informed the structure of much artisanal and peasant culture. Almanacs listed the dates of fairs and markets, and farmers, peasants, merchants and townspeople alike arranged their weekly routines around them. The orchestration of fairs and markets in many areas of France was indeed not just a financial and logistical issue: markets were the great 'moments of sociability', and they were vital to the pride of any self-respecting town or village.[62] Indeed, their importance may have equalled or outstripped that of the Church as a social nexus in many places in France. Hence, markets were integral to, if not the defining charac-

l'exacte observation de l'Annuaire de la République, 23 *fructidor* 6 (9 Sept. 1798), *French Revolution Research Collection*, 6.

58 This is quoted in Thomas, *Les Temps des foires*, 208; 'calendrier républicain', letter to the Administrations centrales, Paris, 29 *germinal* VI (18 Apr. 1798), French Revolution Research Collection, 6.

59 Romme, *Rapport*; Fabre d'Églantine, *Rapport*.

60 On the problem of the number of markets see Young, *Travels*, ii. 757.

61 Villèle, letter to the minister of the interior, July 1806, AN, F^{12} 1274.

62 A.-L. Hartmann, 'Foires et marchés en Maine-et-Loire, 1789–1815', *Annales de Bretagne et des Pays d'Ouest* ci (1994), 61–79.

teristic of, many communities.[63] The introduction of the republican calendar, with its ten-day cycle interrupted these long-established patterns.

According to Jack Thomas's study of the Haute-Garonne, the enforcement of the new scheme led to a 'purgatorial month of *messidor*' in the Year VI (July 1798). The old markets were banned but the new table of fairs and markets remained to be settled, creating enormous problems for local economies, undermining the livelihoods of farmers and merchants, and threatening many households with want, since grain was unavailable for purchase.[64] The timing of the change – at the beginning of the harvest – compounded the problem. Impromptu and illegal markets were held. In Montréjeau, for instance, in the Haute-Garonne, some 4,000 people turned up to a market that the local gendarmes were unable to halt.[65] Like the other legislation on the strict observance of the calendar, the new measures caused disagreements between cantons. The new measures were adopted in Rocheguyon, in the Seine-et-Oise, to the fury of the nearby canton of Vernon, whose merchants complained about the upset to their trade. The town none the less pressed on with their introduction and, in a letter to the interior minister, claimed that they found it necessary to use harsh measures against 'the insubordinates'.[66]

Thomas argues that the correlation between the dates of markets and of notaries' days of business provides an effective method of dating markets during this period. From a study of six towns in the Haute-Garonne he concludes that the new republican markets were a 'marked success between the end of Year VI and *prairial*, Year X'. The majority of markets in these towns followed the new calendar, since they were held on the *quartidi* or *nonidi*, rather than on Wednesday or Saturday. In some places such as Ferran, to the west of Carcassonne, the shift was progressive and gradual, while elsewhere, as in Montréjeau, it could be sudden and marked by violence; but both eventually accepted the reforms. Thomas concludes that 'the change in the calendar as it related to market days had really been accepted by the populations of the Haute-Garonne'.[67] This pattern seems to hold for other areas of France: although the reports contained in the Archives nationales testify to widespread annoyance at the disruption to local affairs, requiring the attentions of the National Guard and repeated reports sent between commune and

63 Ibid.; Thomas, *Les Temps des foires*; P. Lamaison, 'Des Foires et des marchés en Haute-Lozère', *Études rurales* lxxvii–lxxx (1980), 199.
64 For the effects in the south-west of France see Thomas, *Les Temps des foires*, 208–11.
65 Ibid. 208.
66 Canton of La Roche-Guyon, Seine-et Oise, to minister of the interior, 27 *floréal* VI (16 May 1798), AN, F^{12} 1274.
67 Thomas, *Le Temps des foires*.

department, these altercations proved to be something of a temporary problem, and regular markets, following the new calendar, were eventually arranged in many areas.[68]

The change of market arrangements also presented an opportunity for local communities, many of which used the reorganisation of fairs and markets to increase their frequency. Indeed, their number increased by between 10 and 40 per cent. Despite the interior ministry's persistent desire for a reduction in the number of markets, the frequency and number of markets increased during the revolutionary period, as local administrations took advantage of the new regulations to assert their own claims.[69] In the Vendée, fairs were held 'in more cantons than in the old days'.[70] The agronomist Villèle suggested that the number of fairs had doubled compared to during the *ancien régime*, providing one of the republican calendar's more enduring, if unexpected legacies.[71]

Recent historiography has sought to emphasise the continuities of the Revolutionary era, discarding older Marxist certainties of political and social rupture and placing greater stress on longer-term cultural processes, whether secularisation, desacralisation, the rise of the public sphere or centralisation of state control. It is possible, at least in terms of the policing of work, to view the new calendar within the context of growing, and long-term, official interest in the regulation of everyday life and a concern for social order, particularly within towns and cities. Political undertones can be detected in reactions to the new calendar. The *décadi* tended to be ignored in areas of counter-revolution, or in predominately Catholic and rural areas, while in many towns and cities administration and local Revolutionaries gave it ostentatious support. Noah Shusterman's research suggests that reforms sometimes prompted 'dissatisfaction and "murmuring" among certain parts of the population', and even reports of small-scale riots, but never revolt or rebellion.[72] While there are examples of strikes and protest among workers during the Revolution, these were related to wages, rather than the calendar.The scale of reform introduced by the new system was certainly greater than that of attempts at rationalising the ritual year under the *ancien*

68 AN, F^{12}1274.

69 Hartmann, 'Foires et marchés en Maine-et-Loire', 63; Margairaz, *Foires et marchés*.

70 Department of the Loire-Inférieure to the department of the Vendée, 22 *vendémiaire* VII (13 Oct. 1798), AD Loire-Atlantique, L 152.

71 Letter the minister of the interior, July 1806, AN, F^{12} 1274.

72 N. Shusterman, 'Festivals, calendars, and the nationalization of time in France, 1642–1815', unpubl. PhD diss. Berkeley 2004, 12.

régime, but it can be placed on the same continuum; in Toquevilleian terms, it might be described as a centralisation of power, and attempt to bring local variations, such as days of rest, under unified state control.[73] At least this was the aim. Local variation, and sustained resistance to official control, remained common to both the *ancien régime* and the Revolutionary era.

Although calendar regulation may have originated in elite thought about public morality, whether religious or republican, in practice it reflected more mundane concerns about crowds on the street, public drunkenness and the potential dangers of markets. Local officials responded in much the same way to legislation originating from Paris, and enforced it as they saw best. There were still ruptures, or at least profound changes or shifts in the manner or nature of policing. The revolutionary era completed an *ancien régime* project that sought to control time; indeed, it may be said that time was nationalised. The hours of work, days of rest and festivals would be decided on a national basis, and, for the first time, central government took a concerted and lasting interest in the regulation of the calendar. The Restoration placed particular emphasis on Sunday observance by passing a series of laws stating which work or activities could take place on the Sabbath.[74] And, some sixty years later, the Third Republic would confirm the status of national, republican festivals. While the calendar undoubtedly failed as an instrument of social reform and national regeneration, it reveals in greater clarity the instincts of those seeking to create a national calendar, and a coherent system of work and rest. In this way it may be said to presage modern understandings of the week.We now turn to the regulation of smaller measures – that of hours, minutes and seconds.

73 Ibid. 264.

74 *Paris periscope*; Kroen, *Politics and theater*.

6

Republican Hours

Of those members of the committee of public safety arrested along with Robespierre in the summer of Year IV, the youngest and perhaps most ideologically-driven was Saint-Just. He also, it seems, paid a particular attention to his appearance. When he was arrested, he was carrying that most republican of fashion accessories, a decimal timepiece made by A. Elyor, a reputed clockmaker in the Galerie de l'Égalité, the republican name for the Palais Royal.[1] What should we make of this? That the most self-consciously 'Spartan' of all the Revolutionaries clearly saw the need to adopt the new decimal hours of the Year II may indicate the importance that Jacobins placed on the new system. In this politically-charged time, sporting a decimal watch undoubtedly carried a number of meanings beyond mere fashion.[2] For a high-profile member of the committee of public safety, a decimal watch could suggest several things, including his support for rational innovation and the practical benefits of the application of the Enlightenment thought or his rejection of the *ancien régime* and the hours of the Catholic day. The metrification of time also suggested a Cartesian concern for regularity and an interest in clinical exactitude: few would deny that these coldly rational characteristics were present in Saint-Just's pronouncements on the need for Terror or his writings on *republican institutions*.[3] Other well-known republican figures, such as the astronomer Laplace, are also known to have used decimal

1 *The Jewelers' Circular and Horological Review*; Kennedy suggests that such watches were 'extremely rare', although the number of timepieces produced or sold is uncertain: *Cultural history*, 348.

2 The historiography of fashion disagrees on the dating of fashion's birth; less is written on male dress. Clare Haru Crowston argues that 'fashion' began in the seventeenth, rather than the eighteenth century: *Fabricating women: the seamstresses of Old Regime France, 1675–1791*, Durham, NC 2001.

3 Saint-Just considered the festive week in his *Republican institutions*. He made rather strict rules for the observance of the *décadi*, and suggested that it should be a day for the elders to judge the behaviour and character of citizens. He also required citizens in his ideal republic to abstain from eating meat on odd days, although whether this was from stoical or economic motives, or a mixture of the two, is difficult to determine.

timepieces (Laplace's watch was adapted with a new dial). Decimal time is thus easily linked to the extremes of revolutionary culture; this chapter examines whether it had wider consequences, or should be considered a curiosity of the revolutionary era.

Reform of the hour

A calendar is a method of organising days, weeks, months and years. But for many Revolutionaries, the reworking of the measurement of the natural world was not a piecemeal endeavour. Reform of the calendar also allowed for the reshaping of all levels of time, from the *seconde* to the *époque*. Moreover, it was conceived as just one element of wider reforms, ranging from weights and measures to the creation of administrative departments. Such reforms were the natural companions to the political and social innovations introduced by participants in the national assembly and Convention, and the new calendar provided the perfect means of linking them all together.The symbolic framework of a new era, a mapping of republican festivals and a pedagogical programme all found a home in the new calendar, which also handily erased the old religious and royal festivals of what had become known as the *ancien régime*.[4] The new structure of the week, the *décade*, the new names for the months and, not least, the new start and dating of the year announced the scope of the Revolutionaries' intentions. Such reforms can easily be seen as hubristic; indeed, the new calendar largely failed to take deep root beyond the bureaucratic cadre. None the less, the new days and months, as well as the conceit of the republican year, have attracted a certain amount of comment and study by subsequent historians. The republican hours have received less attention. In part, this is due to the general neglect of this aspect of calendar reform by the Revolutionaries themselves. Although the early reports of the committee of public instruction decreed that the days of the year were to be divided into 10 hours, 1,000 minutes and 100,000 seconds, down to 'the very smallest subdivision', decimal time was largely abandoned by 1795 and little administrative time was devoted to it.[5] Few new timepieces were produced, and written traces of how people followed the hours of the day are scarce, at best.

Despite the lack of primary or secondary material and the limited distribution of new timepieces, study of the reform of the hours can still bring into closer focus the links between science and politics, as well as the ways in which the Revolution involved itself in everyday life. While the preceding

4 Baker, *Inventing the Revolution*.
5 Romme, *Rapport*, 5 Oct. 1793, article xi.

chapter examined the contradictory forces involved in the regulation of time, particularly the days on which one should or should not work, the following pages examine how decimal time was linked to a republican concern with supporting modern manufacture and the promotion of the horological arts.[6] The clock or watch, it could be argued, were symbols of the eighteenth century's highest achievements, not just in terms of technological prowess, but also in terms of trade, manufacture and material consumption. There are also cultural and social issues: for example, was the clock primarily a machine for display or was it a tool for the rich to orchestrate their social life? Finally, the shake-up of the hours of the day took place during a period of history during which many have argued that the social meaning of time was undergoing a process of transformation, leading to what Jean de Vries has termed the 'Industrious Revolution'.[7]

Clocks and watches in the eighteenth century

In order to understand the meaning and importance of the republican reform of the hours of the day, it is necessary to characterise what this mode of time meant to contemporaries. Were the hours of the day as important in the late eighteenth century as they are in a modern society, for example; and what cultural or social significance could timekeeping import? There is a case for arguing that time of day had little place in everyday life. In his introduction to the diary of the eighteenth-century artisan and 'proto *sans-culotte*', Jacques-Louis Ménétra, Daniel Roche argues that eighteenth-century temporality differed markedly from modern, time-conscious temporality. Pre-industrial society, Roche suggests, had little need for exact timing or the dominance of the clock, and notes that Ménétra's diary mentions the time of day just thirty-five times in over three hundred pages. On the occasions that the time is recorded, Ménétra used a standard formula of 'Five', 'Six', 'Eight o'clock', 'Ten' and 'Eleven in the evening' as a form of shorthand for points during the day and night. Roche argues that Ménétra's relative indifference to clock time suggests that for the average artisan time was flexible, and from this example he infers that the Parisian popular classes placed little importance on precise timekeeping.[8] In urban workshops, regular hours were less important than the demands of the job in hand, or thirst, hunger and tired-

6 Ibid. ch. v.

7 J. de Vries, 'The Industrial Revolution and the Industrious Revolution', *Journal of Economic History* liv (1994), 249–70.

8 Ménétra, *Journal*, 367.

ness. Unlike the industrial modern age of the factory bell or the office clock, day-to-day life in late eighteenth-century Paris retained a certain fluidity.[9]

This view of pre-modern temporality is most eloquently expressed in E. P. Thompson's influential article 'Time, work-discipline, and industrial capitalism'. Now over forty years old, this important study contrasts pre-industrial people's lack of 'internal clocking' of time with the regular work-discipline demanded by industrialised, modern society: a psychological shift orchestrated and required by the clock and factory.[10] For Thompson, the pre-industrial world did not require sophisticated or, indeed, regular reckoning of time, but was dominated by the seasons and the light of day. While the focus of this study was England, the conclusions may be applied to France, with some caveats about different rates of proto-industrial development. In most of France, after all, piecework was the norm, even though the hourly wage had been introduced in Paris at the beginning of the eighteenth century. The productivity of factories, mills and transport remained at the mercy of the weather and seasons, since a dry stream or river could close a mill or leave barges stranded.[11] In this analysis, clock time is a modern phenomenon, one of the social consequences of the 'Industrial Revolution' rather than a development linked to urban life or the spread of watches and clocks.

But while seemingly commonsensical, this account risks exaggerating the rigidity and dominance of industrial or modern work practices, and also misrepresents the role of timekeeping in pre-industrial society. A number of scholars, particularly historical geographers, have criticised Thompson's depiction of a 'pre-modern time consciousness'.[12] They point to a variety of methods of timekeeping in pre-industrial societies, notably church bells and sundials, as well as emphasising the place of time in agricultural, as well as urban or industrialised, society. If there was a shift in timekeeping, it may well have taken place earlier than Thompson argued. For example, Stuart Sherman has drawn attention to changes during the seventeenth century in the 'outward notation' of time as timepieces became increasingly accurate (in particular a result of the introduction of the minute hand), while Alfred

9 Ibid.; D. Garrioch, *Neighbourhood and community in Paris, 1740–1790*, Cambridge 1986, 170; Sonenscher, 'Work and wages', 166.

10 Thompson, 'Time, work-discipline and industrial capitalism'; Landes, *Revolution in time*. See also Smith, *Mastered by the clock*.

11 R. Whipp, '"A time to every purpose": an essay in time and work', in P. Joyce (ed.), *The historical meanings of work*, Cambridge 1987, 210–36; M. Baulant, 'Salaire des ouvriers du bailment à Paris de 1400 à 1726', *Annales: Economies, Sociétés, Civilizations* xxvi (1971), 465 n. 2.

12 E. P. Thompson, *Customs in common*, London 1991, p. vii. See also Dohrn-Van Rossum, *History of the hour*, 217–89, and Voth, 'Time and work', 29–58.

Crosby has stressed the advances in timekeeping prior to and during the Renaissance.[13] Certainly, by the eighteenth century, as Robert Poole comments, time, 'in the form of clocks and pocket-watches, [was] ever more widely owned'.[14] Recent studies of watch-ownership have suggested that ownership was indeed fairly widespread in French society by the Revolution, and that ownership was growing with some rapidity. It seems likely that as trades became more organised and mechanised over the course of the century the working day started earlier and finished later, as Hans-Joachim Voth has shown by a means of a sophisticated statistical study of English legal cases.[15] Nor was the clock the only mechanism for controlling time: while the working day was often set by the sun, it could also be indicated by the ringing of bells.[16]

Although the importance and nature of timekeeping is difficult to measure with certainty, material culture and textual references can provide some important indications. The prevalence of timepieces provides one clue, but is difficult to quantify this with any great certainty. At the start of the century, watches begin to appear in Paris inventories. Ownership increased rapidly during the century, especially in urban households, and became expected at a certain level of society.[17] Watches were mentioned in nearly all the Parisian clerical inventories examined by Annik Pardailhé-Galabrun, while decorative clocks figured strongly in merchants' inventories from as early as the 1710s. For example, Gilles Curé, a *marchand limonadier*, owned a clock with both minute and hour hands and an enamel dial. This was a high-quality timepiece (at a period when the French watch trade was a world-leader). At the time of his death in 1751, Curé also owned a more valuable silver pocket watch. In comparison, musical instruments make fewer appearances in inventories than watches (this may, of course, relate to their value). Clocks, although more likely to be made of wood and with a leather dial rather than metal and enamel, were also found in more modest households from the beginning of the century.[18] According to this index 70 per cent of servants

13 A. W. Crosby, *The measure of reality: quantification and western society, 1250–1600*, Cambridge 1997; Sherman, *Telling time*, 1–9.

14 R. Poole, '"Give us our eleven days!": calendar reform in eighteenth-century England', *Past & Present* cxlix (1995), 95.

15 Voth, 'Time and work'; L. Fontaine, *History of pedlars in Europe*, Oxford 1996, 193.

16 Slavin, *French Revolution in miniature*, 44.

17 J. de Vries, 'Peasant demand patterns and economic development: Friesland, 1550–1750', in W. N. Parker and E. L. Jones (eds), *European peasants and their markets: essays in agrarian economic history*, Princeton 1975, 205–68, esp. pp. 221–2.

18 A. Pardaihlé-Galabrun, *La Naissance de l'intime: 3000 foyers parisiens, XVIIe–XVIIIe siècles*, Paris 1988, 144, 149, 154, 156.

possessed a watch in the years before the Revolution, while 32 per cent of non-domestic employees owned one: a useful symbol of status and also a useful tool to satisfy the demands of one's employers, who would have set times for meals and a planned social itinerary.[19] Clocks, made cheaply out of wood, may have made mechanical time-keeping much more available to the poorer sections of the population, and their ownership not only gave access to clockwork-controlled hours, but also served to enhance social status. As his journal reveals, Ménétra – as successful artisan – owned a pocket watch. The journalist Louis Sébastien Mercier, surveying the 'new Paris', saw watches everywhere: 'Today, a watch is less an object of fashion than a regulator of commerce. For a people of money, the second hand is the telegraph of the Bourse. If there is a craze for music, a craze for dance, then there is also a craze for watches.'[20] Mercier's impressions, and their large numbers, suggest that their function was not unimportant. Timepieces combined both display and utility, as Saint-Just demonstrated.

Since watches also often attracted the attention of thieves it is possible to use police archives as a source to reveal how people made use of them.[21] The Parisian police archives contain one account by a stonemason who accused a coachman of stealing his pocket-watch. When he hailed the coach he remarked to the driver that it was 'eight in the evening', and that he needed a shot of *eau de vie*. The coachman, the stonemason continued, went drinking with him, but at some point during the evening stole his watch with the help of an accomplice. Although he did not know the name of the manufacturer, the watch had a silver case and was of a 'modern' type, with hour and minute hands and Arabic numerals. The theft was discovered when the victim later attempted to check the time at another bar.[22] His report reveals that it was used several times during an evening's drinking. Why did he need to check his watch during an evening's drinking? The evidence may be unreliable because he had a motive to demonstrate possession to the police, but as well as possibly making social arrangements, or simply being curious as to the hour, he may also have been wondering when the bars may have closed (the Paris authorities attempted to close bars at eleven p.m.; one police account records an argument about the time between the watchman and a drinker).[23]

As this unfortunate report suggests, the ownership of watches and clocks

19 Fontaine, *History of pedlars*, 193.
20 Mercier, *Le Nouveau Paris*, 1197.
21 Thompson, 'Time, work-discipline and industrial capitalism', 69 n. 47.
22 Minutes of the *commissaires de police*, Roule, Paris, 22 *fructidor* XI (9 Sept. 1803), APP, AA 236.
23 AN, Y 9498.

had multiple meanings (again, as Saint-Just demonstrated during the Revolution). Not least of these was a certain pride and social status that could be gained from the sporting of a watch fob. As well as their relationship with fashion and display, their presence brought a more scientific, regulated and precise awareness of time into the world of the everyday and to the streets of Paris.[24] Moreau de Jonnès, a talented but largely luckless artillery officer and future statistician, recalled an encounter with the captain of an American ship *en route* to Martinique. The captain, Jonnès recalled, saw his watch – a 'gold watch with a second hand from Lépine' – and 'his eyes lit up with envy' before he offered to buy the watch. Jonnès, who had been avidly calculating the longitude of the ship with the timepiece, wisely suggested that he would sell it to the captain on his safe arrival. The captain then showed him his collection of eleven watches and explained that his mother, a *gens de couleur libre* who lived on the bluffs of Port-au-Prince in Saint-Domingue:

> nourished me for fifteen years with what she could spare from her poverty. When it was fine, she could tell by the sun the time for work, the only thing which interested us! But when it rained (and it rains half the year on the bluffs) she used to say: 'Ah, if only I had a watch!' On leaving her I swore to bring her one – no, a dozen, and I am not far off the total.

The reader may doubt the veracity of this slightly salty yarn, but it does help to reveal the status that a watch could bring, its cultural significance, and the ambiguous relationship between work and time.[25]

This colonial account is a reminder of life beyond Paris. In contrast to the high levels of watch ownership and related chronological exactitude to be found in the capital, ownership elsewhere was likely to be limited to the elite, particularly in rural areas.[26] Yet lack of a watch did not mean lack of time-consciousness: agricultural workers may have had no need for seconds or minutes, but hours were not without relevance for communal agricultural work since plough teams and so forth often relied on coordination between sets of labourers who might meet in a field at a certain time. Fishermen also had to be aware of the hour and day to comprehend the tides, as did those making calculations about the setting of the moon when hunting. Time could be told in a number of ways, such as by bells or sundials, or discerned from the sound of traffic or the height of the sun. Timekeeping was not just

24 Pardailhé-Galabrun, *La Naissance de l'intime*, 161.

25 A. Moreau de Jonnès, *Adventures in the Revolution and under the Consulate*, London 1929, 218.

26 R. Lick, 'Les Intérieurs domestiques dans la seconde moitié du XVIIIe siècle, d'après les inventaires après décès de Coutances', *Annales de Normandie* iv (1970), 308.

entrusted to the hands and dials of pocket watches, recently developed carriage clocks and more imposing *horloges* and *pendules*, but was more traditionally displayed by sundials, which were found on many public buildings, churches, or found by using portable gnomons.[27]

Timekeeping also played a role in France's financial and transport networks. For those involved in business, time played a significant role: an accurate and common system of timekeeping was necessary to all but the most basic financial or business transactions. The agenda and date book entered into common use during this period, and financial operations, in particular credit agreements, were based upon calculations of time: interest accrued by the day. Almanacs – which were produced for all sections of society from farmers to the army, and clerics to courtiers – contained a wealth of temporal information. The timetables within them reveal a surprising degree of chronological accuracy, and concern with hours as well as days and months. Coach, canal and river travel times suggested that the vehicles left at the time indicated, as the advertisement warned, 'on-the-dot'. Mail times were particularly punctilious, at least in principle. Finally, meal-times provided a further chronic marker. Eating provided one of the set points of the day, as the English traveller Arthur Young reveals in a complaint about the fondness of the French for midday dining among most classes:

> This single circumstance, if adhered to, would be sufficient to destroy any pursuits, except the most frivolous. Dividing the day exactly in halves destroys it for any expedition, inquiry, business demands seven or eight hours' attention, uninterrupted by any calls to the table or the toilette [...] I am obliged to make this observation, because the noon dinners are customary all over France, except by persons of considerable fashion in Paris.[28]

Untroubled by the perils of midday dining, high society was still, to some extent, governed by the clock, no doubt in part due to the influence of the regular routine of the king and court at Versailles. For example, the *Almanach nécessaire* for 1789 provided extensive listings of *bals de l'Opéra* on Sunday nights, noting that they cost 6 *livres* and began at midnight.[29] Even officials in what would become defined as the *ancien régime* had to keep office hours: Monsieur de Barentin, the Lord Chancellor held court Sundays from 12 p.m.

27 H. Monod-Casidy, 'Un Astronome-philosophe, Jérôme de Lalande', in T. Besterman (ed.), *Studies on Voltaire and the eighteenth century*, lvi, Geneva 1967, 914–15; S. Tardieu, 'La Vie domestique dans le Mâconnais rural préindustriel', unpubl. PhD diss. Paris 1964, 609.
28 Young, *Travels in France*, 33. On French mealtimes see A. Corbin, *Le Temps, le désir et l'horreur: essais sur le dix-neuvième siècle*, Paris 1990, 9.
29 *Almanach nécessaire*.

until 2 p.m. at Versailles, while more usual *jours des audiences* were arranged between 10 a.m. and 2 p.m. most days of the week.[30]

Important as the Lord Chancellor may have been, produce and livestock markets played a vastly more important role in the life of the country. These institutions also had a particularly close – and ancient – relationship with timekeeping.[31] This intimate connection was longstanding: by the fourteenth century, markets began to be regulated by the hour: sales of salmon, for example, were prohibited before 6 a.m. in summer and 8 a.m. in the winter; wholesalers could only be served on Fridays after 10 a.m.[32] Similar statutes were still in place at markets throughout the country five hundred years later.[33] The authorities regularly policed the opening times for markets, which, judging by the ordinances, the traders often attempted to flout by opening earlier than was allowed by the rules.[34] The *Ordonnances et sentences de Police du Châtelet de Paris* regularly detailed the hours of trade at markets, revealing very early starting times.[35] Typically, the hours during which fruit could be sold were regulated so that ordinary people could buy what they needed before wholesalers bought all the stock. Neglect of such regulations could lead to small-scale riots.[36] Markets were always a potential problem for civil order, and the correct temporal limits helped to contain this problem; indeed it was one of the few practical methods of keeping order and providing markets with some form of just regulation.[37] The site of clocks in towns, usually within view of the market, greatly contributed to this coordination, and to a large part explained their existence.

Thus the hours of the day played an important role in many people's lives, from domestic servants to the elite. Although an exact counting of the hours meant little to many, particularly in the countryside, few were unaware of the passing of time, particularly as a result of the ubiquity of bells. Nor was this pattern unchanging, but over the course of the eighteenth century, awareness of time was undoubtedly heightened: watches and clocks became much more affordable, the development of regular postal and coach routes not only

30 Ibid.

31 G. W. Skinner, 'Marketing and social structure in rural China', *Journal of Asian Studies* xxiv (1964), 11–16.

32 Dohrn-Van Rossum, *History of the hour*, 234.

33 Hartmann, 'Foires et marchés en Maine-et-Loire, 61–79, esp. p. 68.

34 H. Hauser, *Travailleurs et marchands dans l'ancienne France*, Paris 1929; Dohrn-Van Rossum, *History of the hour*.

35 AN, Y 9498–9.

36 Extrait du registre des délibérations de la municipalité, 4 Aug. 1809, AM Cholet, cited ibid.

37 Dohrn-Van Rossum, *History of the hour*, 249–50.

began to shrink the country geographically, but also made it more temporally uniform, while work practices began to be more aware of work-time, rather than piece work. The great publishing event of the age, the *Encyclopédie*, also placed a great deal of importance on the art and science of horology, speaking of an intellectual concern with the exact measurement of the day. And, as the regulation of markets showed, time was intimately connected with the ordering of society, something that resonated during a revolutionary era.

Reform of the hour and decimal time

Some months after the new calendar had been introduced, the Convention arranged for a competition to promote the new form of hours. In *fructidor* II (April 1794), a *jury au nouveau système horaire* convened to judge the suggestions for ways of adapting timepieces to the new manner of timekeeping. Appropriately enough, it met in the Salle du Bureau de Consultation des Arts & Métiers in the Louvre and counted amongst its membership the famed clockmaker Ferdinand Berthoud and the mathematician and astronomer Charles Lagrange. In all, the committee examined 191 plans for a variety of decimal timepieces. The majority, judging by the scathing reports of the jury, were not of a high standard, but eventually they managed to whittle the designs down to a shortlist. A clockmaker from Carouges, near Geneva, claimed victory and 1,500 *francs*.[38]

The committee's deliberations point to the uses to which clocks and watches were put and the importance that was placed on the new division of the day. Firstly, horology was 'an art as equally useful to society as to commerce'; as an industry, it was important to everyday life, as well as to the economy.[39] The jury also distinguished between the needs of science and of the public. As a result they divided the competition into two: one for highly accurate scientific instruments suitable for calculating longitude; and the other for personal watches and public clocks. The latter, it was recognised, were 'machines of the greatest importance for the use of all citizens'. The committee faced the problem of the impact that reforms would have on the watch and clock trade. It was feared that the loss of foreign markets would be a financial disaster for the already struggling, but prestigious, French watch trade.[40] National pride was at risk: a nation based on reason and proud of its technical abilities should be able to boast superiority in these arts.

38 *Rapport sur les questions relatives au nouveau système horaire*, Paris Year IV (AN, F^{17}1135 85).
39 Ibid. 4.
40 Ibid. 42.

Figure 8. Silver cased verge watch with decimal dial,
Cuenin, Besaneon [1795–1805]
(British Museum, 1958, 1201.832)

Decimal time was not limited to Paris, but was enforced with differing degrees of enthusiasm among the new departments. Officials in Dijon, for example, were told that in the Côte-d'Or acts had to be dated and time-stamped according to the new hour and were given a paper *cadran comparative* to convert the time. A clerk in Montélimar in the Drôme recorded the sale of *biens nationaux* as taking place on the 20 *germinal* (9 April) at 'three hours, seven *décemes*, modern time', corresponding to nine hours in the morning old-style'.[41] Clearly, the traditional hours of the day were being used in practice, even if it was then also recorded and transcribed in the new fashion. The minutes of some meetings in the Doubs appear to have used the new time and the civic administration in Marseilles used decimal hours for recording the minutes of their meetings, while in Toulouse, several Jacobins sported decimal timepieces.[42] A local watchmaker had been arrested on suspicion of

41 J. Nicolas, *La Révolution Franeaise dans les Alpes-Dauphiné et Savoie, 1789–1799*, Toulouse 1989, 256.
42 Department of the Côte-d'Or to the municipality of Dijon, 29 *pluviôse* III (17 Feb. 1795), AD Côte-d'Or, L 1701; Kennedy, *Cultural history*, 348.

holding royalist opinions, but the *comité de surveillance de Toulouse* released him so that he could construct decimal watches.[43] The municipal administration also decided that the clock on the Capitole should be altered for the benefit of Toulouse's citizens. The alterations took four months to complete, and although the bells could not be altered (and would ring the old hours), the hour hand could (there was no minute hand), and for five years (Years III–VIII), Toulousains could take the time from a decimal dial on the front of the Capitole.[44] The Haute-Garonne departmental administration also ordered that the decree on the new measure be reprinted and distributed amongst the principal clockmakers.

Private enterprise also played its part. In the Year II, the *Courrier patriotique* announced the sale of decimal clocks made in Névache in the Brianeonnais. These devices proudly displayed 'the new division of time', as well as the old hours. The clocks sounded quarter and half hours, and they would be useful in announcing 'popular meetings'.[45] As was noted earlier, *Les Décades républicaines* advertised a paper decimal 'montre économique'.[46] Such 'concordances' point to the difficulty of making mechanical adjustments to existing timepieces. Although it was a relatively simple matter to replace the dial of a watch or clock to show five hours rather than twelve, the sweep of the minute or second hand required adjustment (though it should be pointed out that many clocks just had an hour hand). Timepieces with alarms or bells also posed special technical challenges since without alteration they would sound the old hours. More expensive clocks and watches sometimes incorporated complex calendar mechanisms or automata that would have required complex alteration. Without modifications to the timepiece's innards, clockwork-driven bells would not sound correctly on the hour or quarter hour. These difficulties in part account for the tardiness in converting the clock overlooking the National Convention in Paris. Modifications were first suggested on 17 *frimaire* II (7 December 1793), but the clock, which had four different faces and also worked as an astronomical calendar, proved difficult to alter. It took several years, and eventually two faces were decimalised; one contained a dial for the republican calendar year.[47]

43 Rey-Pailhade, 'Calendrier républicain', 451.

44 Ibid. 444, 54.

45 Nicolas, *La Révolution Francaise*, 256.

46 A. Sérieys (ed.), *Introduction aux décades républicaines, ou à l'histoire abrégée de la république francaise, etc.*, Paris Year II (1794), 8 *pluviôse* II.

47 'Pour tous les travaux faits à la grande horloge à Equation sonnant les heures et les quarts, marquant les heures et les minutes sur quatre grands cadrans, dont deux indiquent le tems en division décimale, placés dans le pavillon de l'unité du Palais National du Conseil des Anciens': Hurcault, Report on the clock for the *Corps Législatif*, Year VII, AN, C 445.

The birth of the Besançon watch industry

As J. P. Gross, Ken Alder and others have shown, the Jacobin authorities endeavoured to introduce a range of national economic reforms. As a result, a series of schemes for national 'factories' were proposed. In part these were a desperate, and ultimately successful, attempt to marshal materiel for the nation's defence, in part they were a component of the Jacobins' economic and social project.[48] For example, as part of the Jacobins' welfare programme, it was proposed that factories should be set up to provide work; and the nationalised naval yards were vigorously organised for the republic's purpose. Although these schemes fitted with the Jacobin need to organise the nation's resources towards a military purpose, they also derived from pre-revolutionary economic thinking which sought to direct industrial output towards the aims of the state. By the end of 1793 there was a danger that such schemes might be associated with the overthrown Girondin faction. For example, Étienne Clavière and Étienne Dumont had promoted a scheme for transplanted watchmakers in Ireland, dubbed 'New Geneva' in the early 1780s, but the potential economic and political gains proved irresistibly attractive to the Jacobins.[49]

French watchmaking had been an important industry, particularly at the luxury end of the market. The industry specialised in fine clocks and watches, with elaborate enamelled dials and ornate cases fabricated or decorated in gold. Because of the patronage of the king and the court, France had also been able to boast of the presence of some of the most able and innovative clockmakers. In particular, in the 1760s and '70s Pierre LeRoy and Ferdinand Berthoud had placed France at the forefront of chronometry, while, on the other hand, the market for cheaply-produced clocks and less advanced clockwork innards had become dominated by artisans in the Swiss cantons. The disruption to the trade in luxury goods and the economic dislocation caused by the war presented the industry in the early 1790s with a potential crisis. An enterprising Swiss horologer, Laurent Mégevand, took advantage of this opportunity and managed to persuade the Jacobin authorities to support a national watchmaking factory in Besançon, in the Doubs. By 1794, between two and three thousand Swiss watchmakers had emigrated to the town, forming the basis for the nineteenth-century French mass watchmaking industry.[50]

48 Gross, *Fair shares for all*; Alder, *Engineering the revolution*.

49 The Irish parliament had attempted to create a 'new Geneva' with the support of £200,000 for a manufacture: *Rapport fait par Lemercier*, Paris Year VI, AN, F^{12}1326.

50 F. Arnould, 'Laurent Mégevand et l'introduction de l'horlogerie à Besançon, 1793–1814', MA diss. Besançon 1989; N. Petiteau, *L'Horlogerie des bourgeois bonquérants:*

Mégevand was born in Geneva on 11 January 1754, the youngest son of a well-to-do clockmaking family with French origins. He followed his father into the trade and became a master clockmaker by the age of twenty-three. His early career reveals his somewhat cavalier spirit of enterprise. On two occasions the city's gild of clockmakers found him guilty of attempting to pass off eighteen-carat gold as a purer grade of metal. On the first occasion he was fined and placed under the supervision of his father, who failed to keep a close enough watch on his son, since in 1778 Mégevand was again convicted. Now, the court expelled him from the city. Mégevand left Geneva and travelled north, to Le Locle. Here he married and, with his father-in-law's financial help, set up in business, entering into partnership with a local clockmaker. Debts soon forced the closure of the company, but by 1791 Mégevand's new business, now as a watch trader (*négociant*), rather than as a manufacturer, appears to have been flourishing. He now had international connections and travelled on business to Paris and Amsterdam and made trade links with Bordeaux, Lille, Lyons, Saxony and Russia. During this time he also made contact with the French banker Étienne Clavière, and discussed the possibility of establishing a watchmaking colony to avoid French import duties.[51]

By the end of 1792 news of the French Revolution had created a revolutionary spirit among many in Le Locle and La Chaux-de-Fonds. Liberty trees were planted, popular societies formed and republican cockades worn. In response, in April 1793, the Prussian authorities expelled around 200 people, the majority of whom were watchmakers, and many of whom headed across the French border. Mégevand was among those exiled, and the watchmaking colony took on a new urgency.[52] Dijon, where Mégevand had previous business connections, was first considered as a base, but, according to Mégevand, the wives of the artisans thought that this would be too far from their homeland.[53] Besancon in the Doubs offered a number of advantages, in particular

histoire des établissements bourgeois de Damprichard Doubs, 1780–1939, Besancon 1994; J.-L. Mayaud, *Besancon horloger, 1793–1914*, Besancon 1994.

51 E. Clavière (1735–93) was involved in the revolt of 1782 before making his fortune in banking and speculation in France and overseas. He became Finance Minister in March 1792 as a result of Brissot's patronage, but was eventually imprisoned for his Girondist associations in June 1793, and committed suicide in his cell; L.-N. Lemercier, *Rapport fait par Lemercier, Député de la Charente-Inférieure au nom d'une commission spéciale, composée des représentants du peuple Bordas, Missonet & Lemercier, sur la résolution du premier nivôse an 6, relative aux citoyens Mégevand & Trot, chefs de la manufacture nationale d'horlogerie de Besancon: séance du 14 messidor an* 6, Paris Year VI (AN, F^{12}1326).

52 He may also have been in trouble with the authorities for the way in which he conducted his business.

53 Report by Mégevand, 7 Sept. 1807, AN, F^{12} 2434. Mégevand reflected that Dijon

its location on trade routes between the Rhine and Saône and as a transit town for merchants travelling between Geneva and Basle, or Neuchâtel and Paris; furthermore, it was already an important node in the watch trade, although it lacked any horological manufacture.[54] As a predominately Jacobin town, surrounded by a deeply Catholic rural population largely hostile to the Revolution, the influx of pro-revolutionary artisans had its attractions to the members of the local Jacobin club. Pierre-Joseph Briot, the editor of the town's paper, *La Vedette*, and a prominent Besaneon Revolutionary, was also a strong supporter of the project, not least because an influx of political radicals would help to shore up the city's Jacobin support.[55] Petitions were printed extolling the virtues of horological manufacture. Jean-Marie Calès, for example, attempted to persuade the citizens of Besaneon that this influx was a benign development, bringing a new and profitable industry to a mountainous region of 'sterile rocks'. Horology was an art which could be practised by all, 'including females' and both old and young.[56] He argued that 'we can no longer make our children into [...] priests, monks or financiers [...] It is necessary to give them as an inheritance an interesting and useful art; and this is horology'.[57]

The colony became a Jacobin project, run by the committee of public safety 'under the protection of the republic', and was subsidised to the tune of 250,000 *francs*. The local ordinances decreed that, 'those that would halt its progress, or bring about its decline, must be pursued as troublemakers and foreign agents'.[58] Eventually some 850 migrants moved to Besaneon from Le Locle (18 per cent of the Le Locle population). By the Year IV, 1,855 people had arrived in the town, the majority of whom came from the principality of Neuchâtel; the remainder originated from Geneva or the bishopric of Basle. It was a significant influx; the immigrants constituted some 8 per cent of Besaneon's population of 22,000.[59] This initial immigration was followed by other groups of watchmakers, in particular Jean-Franeois Auzière, whose manufacture employed fifty artisans and their families (making about 200 people in all). There were teething troubles. French Bisontins were aggrieved

would have proved the better choice, as in Besaneon they were subject to 'religious and political fanaticism'.

54 Arnould, *Laurent Mégevand*, 94.

55 Mayaud, *Besaneon horloger*, 14.

56 J.-M. Calès, *Proclamation du représentant du peuple J. Marie Calès, délègue par la Convention nationale dans le département du Doubs, pour y protéger l'établissement de la manufacture d'horlogerie, aux citoyens de la commune de Besaneon*, Besaneon Year III.

57 *La Vedette*, 3 *prairial* II (22 May 1794).

58 Decree, 13 *prairial* II (1 June 1794), AD Doubs, L 369.

59 Mayaud, *Besaneon horloger*, 17.

by the newcomers' exemption from military service and by the financial assistance that they had received from the Jacobins. These complaints were not unique. A similar influx of outsiders also caused difficulties at the Versailles *horlogerie automatique*, as local people complained of apprentices stealing lead.[60]

Despite the influx of skilled workers, the colony proved an early financial failure. Poor trading conditions resulting from war and the dislocations of the Revolution, the rapid decline in the value of the silver *marcs* provided by the government, the difficulty of buying expensive raw materials with *assignats* and over-investment in property and plant all played their part in undermining the profitability of the early enterprises.[61] In total, Mégevand produced only 17,675 timepieces, fewer than anticipated. He later sought favour under Napoleon, but without success. After a period in Paris, the entrepreneur returned to Besançon a ruined man where, in 1814, he was killed during the Austrian siege of the town. Other firms failed to find the success that they had hoped for: Auzière and his workers produced just 1,218 timepieces by his death in 1798.[62]

Despite the early disappointments of these enterprises, the innovative work practices that they introduced have been seen by some historians as harbingers of a more industrial and regulated era, as well as forming a pool of talented artisans who would assist in the creation of the key nineteenth-century French watchmaking firms in the area. It is possible to see the new enterprises as engendering Thompson's 'internal clocking'. The three main Besançon employers (Mégevand, Trot and Auzière) all provided accommodation for their workers, enabling them to supervise many aspects of their lives. Production was broken down into smaller tasks and was closely regulated. Such organisation was very different from the traditional watch- and clockmaking of the mountain regions, which was typically an activity for the winter months, undertaken on a small scale by artisans who also farmed during the summer. In contrast the large and successful Besançon factories that grew out of the first national manufactures, such as those of Japy, Chaffoy, Girod de Naisay and Beaupré, produced watches on an industrial rather than an artisanal scale. By 1810 around 17,000 timepieces were being produced annually in Besançon.[63]

60 Report on the 'horlogerie automatique de Versailles', *Bureau des arts* to the Minister of the Interior, 11 *fructidor* XI (29 Aug. 1803), AN, F^{12}1326.
61 Mayaud, *Besançon horloger*, 22–4.
62 *Acte de décès d'Auzière*, 2 *floréal* VI (21 Apr. 1798), AD Doubs, L368.
63 Arould, *Laurent Mégevand*, 190. True mass production methods were to be found shortly after in the United States, in particular at Eli Terry's factory in Plymouth, CN: Landes, *Revolution in time*, 310–11.

Work discipline

E. P. Thompson's characterisation of work discipline as a modern, industrial characteristic needs to be qualified by the extent to which temporal signals, such as bells and sundials, as well as the widespread use of timepieces had already provided a means of the 'internal clocking of time'. It is also easy to exaggerate the regularity of industry, and the ways in which workers maintained control of their own time. However, the concept of work discipline still has its value, not least in drawing attention to the ways in which time and control were linked in the past, and how this social behaviour waxed and waned over time.

The Bisontin authorities were acutely aware of the links between order and the clock. Charles Dupuis, the government representative at the national manufacture, reported to Paris that the town suffered from thieves and rebellion. In particular, apprentices were proving to be a more explicit discipline problem. In an attempt to control the situation, Dupuis drew up a set of regulations that relied on stringency and a sense of temporal order. Under a strict code, the clock would sound at 5 a.m. in the summer and at 6 a.m. in winter. If the apprentices were not ready within a quarter of an hour, then they were to be fined 1 *sol* for every fifteen minutes that they were late. Mealtimes were similarly strictly regulated. Breakfast was scheduled for a quarter of an hour at 8 a.m., lunch was at 12:30 p.m. and supper was to be eaten between 6:30 p.m. and 8 p.m. Apprentices were to be in bed by 9 p.m., when the doors were locked.[64] Any infractions of these regulations were to be announced publicly at the next *fête décadaire*. Discipline was to be internalised through regulation and public denunciation. Social shaming as well as financial discipline inculcated temporal regulation and routine. The watch factory, in this vision, resembled the clockwork of the end product.

These social mechanisms were not limited to the factory. As ever, Michel Foucault is both suggestive and contentious on this question of timing, power and control, emphasising in particular the importance of school:

> Time penetrates the body and with it all the meticulous control of power [...] The various chronological senses that discipline must combine to form a composite time are also pieces of machinery [and] it was probably in primary education that this adjustment of different chronologies was to be carried out with the most subtlety [...] The school became a machine for learning.[65]

64 Report, Charles Dupuis, Besancon, to the minister of the interior, 2 *fructidor* V (19 Aug. 1797), AN, F^{12}2403.

65 M. Foucault, *Discipline and punish: the birth of the prison*, London 1977, 164–5.

Prescription similar to Dupuis's measures can be seen in proposed school timetables. Many similar codes survive in departmental archives and demonstrate the republican concern with work and order. Syllabi and curricula were carefully worked out, and the school day carefully timetabled. But Foucault also highlights the difference between what might be termed Cartesian prescription and the ragged reality of practice. In reality, as R. R. Palmer noted, teachers could vary the hours to suit; enforcing a revolutionary calendar proved difficult at best.[66] In Besanèon complaints about the apprentices' behaviour continued to be made, suggesting that Dupuis's regime was not always followed. Temporality by necessity had a certain flexibility, avoiding neat divisions into arbitrary hours. Seasonal variations affected the working day and the periods for which men and women were available. Harvest time naturally dictated when people needed to be free for other activities, while winter snows could cut off mountain villages. Manufactures and factories also depended on the season for their power, as mills relied on a steady stream of water, which could dry up in summer, or become an unusable torrent at times of storm or flooding. Proposed timetables built in flexibility by suggesting that students could study subjects at different times, rather than being forced to follow a strict programme.[67] The changing hours of sunrise and sunset throughout the year affected the hours of schools, bureaux and manufactures, with many altering their hours according to summer or wintertime in an early form of Daylight Saving Hours. Time was still too closely associated in people's minds with the sun's passage for it to become based on what could be read from the dials of a mechanised watch or clock. Solar time, rather than the clock time termed 'mean time', predominated.

The Revolution, however, may have given a further impetus to the regularisation of time. Despite the fluidity of solar hours, interest in natural science and public utility underpinned a concern for exact timekeeping and efficient use of the hours of the day. The Revolutionaries had, in James Leith's phrase, 'an obsession with uniformity', which extended to timekeeping.[68] Commentators noted the importance of time and exact awareness of it for industry. Mercier deplored the lack of public clocks in Paris, because 'the fracas from the coaches' made it impossible to detect the 'small chimes' of carillons that told the hour. He suggested that the public magistrates ought to arrange for clocks to be erected in the *faubourgs* where there were many artisans as a help and encouragement in their long labours.[69]

66 R. R. Palmer, *The improvement of humanity: education and the French Revolution*, Princeton 1985, 245.
67 AN, $F^{17}1338$.
68 Leith, *Space and revolution*, 216.
69 Mercier, *Le Nouveau Paris*, 1307.

The more exact and regulated timekeeping was magnified by the dangers facing the republic. The material shortages of the Years II and III and the demands created by war led to exhortations to greater work and calls to squeeze the most work out of the day as possible. For example, in March 1794, the Paris arms workshops began to work by candlelight, adding two hours to the day without a related rise in wages.[70] The national shipyards similarly enforced strict work patterns, where supreme effort was required to satisfy the needs of the navy following the battle of Brest. On 15 *pluviôse* II (3 February 1794), the committee of public safety decided to increase the number of hours worked. Work was to commence at five in the morning and continue until 7 p.m. in summer or until dusk in winter, although some ports such as St Malo had slightly different hours, where work started half an hour earlier and continued later. Half an hour was set aside for lunch, and announced by the ringing of a bell at the start and end of the break. The ten-day week of the republican calendar reduced the amount of time off, since Sundays (and the drunken furloughs of the *jour de saint lundi*) were proscribed. Government representatives at Rochefort wrote that no workmen had taken the day off on the previous *décadi*, recognising rather grimly that they should only stop work when 'we have no more enemies to exterminate'.[71] The representative claimed that work continued for twenty-nine days consecutively, although the thirtieth day had been used to pay the workers. Perhaps not coincidentally, overtime was paid at a better rate. The sources do not reveal for how long this practice continued, although as there are no further mentions of continuous working, it seems likely that a slightly less exhausting routine took its place. However, under Jeanbon Saint-André's labour code, workers' time was strictly regulated amongst the nation's shipbuilders and repairers. While the republican elite may have believed in piece-work, which, in Prieur's words, 'give the worker a better sense of dignity' and fitted the Rousseau-inspired image of the self-reliant artisan, in practice, a daily wage had to be paid. Such codes were typical of the period, with strict punishments for the idle, and rigorous control of time. Administrators of arms manufactories, for example, 'deployed the rhetoric of revolutionary fraternity, praising the patriotism of their comrade-workers, invoking the higher national purpose'.[72]

Despite their importance for the national war effort, it should be remem-

70 Alder, *Engineering the Revolution*, 253.

71 AN, B.B. 3 60, fo. 210, quoted in N. Hampson, 'Les Ouvriers des arsenaux de la marine au cours de la Révolution Francaise 1789–1794', *Revue d'histoire économique et sociale* xxxix (1961), 269–329, 442–73, quotation at p. 447.

72 Ibid.; Alder, *Engineering the Revolution*, 269.

bered that such national establishments were the exception to usual patterns of work. Extreme efforts were required for the war effort, and the yards worked under direct government authority, with a martial code. Still, a concern for maximising labour and control of time can also be detected in less vital industries. To some extent, this was something of a moral question that predated the Revolution. A crusade against laziness and idleness was underway in the religious pamphlets of the period, which drew on a long tradition of moral condemnation of idleness. At the same time, most thinkers and moralists had long placed great importance on work and its relationship to the wider society.[73] Colbert had related work to the happiness of society or, to be more exact, the wretchedness of people to lack of work. Poverty and economic well-being are clearly central concerns, but the decision to work also created a secure and comfortable society. 'The poverty of the people of the towns and provinces', Colbert suggested, 'does not lie in the taxation that they pay to the king, but only in the difference that there is between the work of the people of one province and another, because they are comfortably off from the moment that they want to work.'[74] Seen as such, work was a social question, a view even more clearly seen by the time of the Revolution (although misery would have been put more squarely at the door of the king). Work in the name of the *patrie* was clearly a virtue, and one to be celebrated and enforced by revolutionary festivals (such as the festival of Labour). Administrators in Amiens believed that the republican calendar would 'rekindle' a love of work.[75]

As well as a devotion to duty and nation, the profit motive was also in force when it came to those Jacobins who controlled industries such as the proto-factories in Besaneon. As Dupuis's code shows, production was tightly controlled in the Besaneon manufactures. *Comptes décadaires*, following the republican calendar, had to be completed by the artisans and their overseers. For example, one machine operator informed Beuque, his controller, that he had 'stamped four hundred and twenty-five watch mounts of different sorts from the 10th *brumaire* to the 20th, Year III'.[76] As this example suggests, production was also specialised to a degree, with some artisans concentrating on dial enamelling, others on the fabrication of clockwork components, and others assembling the final pieces. Children and women were also employed in a number of these stages of production, ensuring that an industrialised sense of time was not limited to grown men.

73 Jacob, *Le Travail.*

74 Quoted ibid. 31.

75 *Délibération de l'administration municipale d'Amiens*, 13 *thermidor* VI (31 July 1798), AD Somme, L 421.

76 *États décadaires*, Year IV, AD Doubs, L 377.

Figure 9. Copper single year calendar medal, Paris 1797
(British Museum, 1901, 1115.2)

In many ways, the Revolutionary era hastened the spread of what may be termed modern, industrialised time, shaking France loose from traditional structures and time-honoured practices. The demands of war and the administrative power of those running organisations such as the national shipyards or watch manufactories, provided the opportunity to regulate working hours and instil an ever-present sense of time. To some extent the end of gild regulations paved the way for other forms of production. It is also tempting to see the military, with its insistent focus on drill and order, as another motor of increased time-awareness.Yet against these examples (and republican aspirations of a virtuous workplace) must be placed the vast disruptions of the period; the collapse of many industries and trades, such as the luxury goods manufactures of Paris and Versailles, and financial dislocation, removed traditional structures of time. Military service, counter-revolution and the Terror shattered traditional routines, and the suppression of religious services and ineffective take-over of religious hospitals and schools dislodged some of the ways time was imposed on society. The republic's war on the 'external signs of religion' and the military's hunger for metal destroyed the aural map of time that France's great network of bells provided.[77] France also remained an agricultural nation, following a way of life linked to the seasons of the year more than the hours of the day. Those changes in the importance of temporal

77 A. Corbin, *Village bells: sound and meaning in the 19th-century French countryside*, New York 1998.

regulation that did take place arose as much out of ideological and cultural demands as economic shifts. The codes of the shipyard and manufactory owed perhaps as much to conceptions of duty, nation and reason, as they did to capitalism.

However important the reforms of time had initially appeared, by the end of the Year IV decimal time had been sidelined by the Convention; nor was it resurrected by the Directory, despite the importance that they placed on the republican calendar in later years.[78] The material record also points to decimal time's slight penetration. In Toulouse at least, the number of tables of concordances between the old and new systems printed was too small to have any impact on the general public. In Paris, the number of such concordances published was greater, yet one suspects that for the majority of citizens decimal time was an unnoticed novelty.[79] This said, decimal clocks did not vanish, even after decimal time was sidelined by the Convention. In 1798, for example, Thomas Bugge, a Dutch scientist who had been invited to France to work on the new measures, noted that he had come across two clocks divided 'according to the new time' during a visit to Paris.[80] While the decimalisation of time linked neatly to the rational designs of the Revolutionaries and the republic's interest in work, such a transformation lacked any real political backing or practical purpose. It allowed the state to interfere with everyday life in an attempt to match the quotidian with an abstract ideal – but such interference was little more than dabbling. It stands as a testament to both the reach of the Revolutionaries' utopian vision and its often unrealisable nature.

78 *Procès-verbaux*, ii. 894.

79 Rey-Pailhade, 'Calendrier républicain', 448.

80 T. Bugge, *Travels in the French republic: containing a circumstantial view of the present state of learning, the arts, learned societies, manners &c. in that country*, London 1801; C. Cardinal (ed.), *La Révolution dans la mesure du temps: calendrier républicain heure décimale, 1793–1805*, La Chaux-de-Fonds 1989, 71; Klein, *The science of measurement*, 124.

Conclusion: The Legacy of the Republican Calendar

This book has sought to uncover the history of revolutionary time not just in the pages of almanacs, diaries and official pronouncements, but in the ways that the years following the dramatic events of 1789 shaped the experience of everyday life. Accounts of the reshaping of everyday life have usually been stories of conflict or negotiation, often framed by intensely local contexts, but given political and ideological colouring. In contrast to the great *journées* of 1789, such as the storming of the Bastille or the abolition of feudalism, they may also appear at first glance to be matters of relative insignificance: the date of a meeting; the smudged pages of an almanac; the periodicity of the filing of official reports. Unlike the tearing up of the *ancien régime* on the night of 4 August and the days that followed, the reworking of everyday life leaves less obvious traces. None the less, the ordinary and the quotidian had their own inescapable importance and many of the Revolutionaries and their opponents understood the potential, and even the necessity, of planting the Revolution's flag on this terrain. It was in such details that the great claims for human progress of the Enlightenment and the Revolution could be mapped on an intimate scale and might make the leap between abstract schemes and the reality of a new, regenerated nation. The care that Romme took over the *Annuaire du cultivateur*, his patterning of the decade, month and year with the instruments, creatures and plants of agriculture and animal husbandry, suggests that he saw the cultivation of the land as the basis for a just and productive society. Such agricultural rhetoric may not soar today in the same way as the Declaration of the Rights of Man and the Citizen still does, but it is eloquent in its own way about the desire to reshape society and the belief that the soil or the vine might be the garden for a new citizenry.

At the same time, we cannot escape the fact that the calendar was a failure; for most of the years of its existence, we can detect an awkwardness and even an embarrassment about the artificiality of the new system, its continued coexistence with the old, forbidden calendar, and its contrast with the calendar used by the majority of other nations. Its existence, and widespread disuse, was a constant reminder that the aims of the Revolution had not yet been achieved and an admission that the republic was, at best, a work in progress. As Sanja Perovic has argued, it was a sign of change, but also a sign of the impossibility or the failure of that change.[1] With these difficulties, the calendar may be interpreted alongside other republican projects, such as the creation of a national administrative cadre and a functioning system of national government, the beginnings of a national educational programme, and the constantly revised system of national festivals that sought to embody and inculcate national feeling and identity. This book has largely concentrated on the extent to which the republican calendar – and the sense of revolutionary time – may be placed in a longer-term context, examining how, for example, markets and the regulation of work might be related to projects under the *ancien régime* to regulate morality and public space. In these final pages I would like to reverse the temporal focus, looking forward to what the republic's efforts to impose a single, national, and secular time on France may have meant for the years and decades that followed.

The calendar abroad

As well as a chronological legacy, the calendar expanded geographically as republican time followed in the footsteps of the Revolutionary armies. As republics sprouted across Europe aided by columns of French troops, a series of 'first years of liberty' needed to be recorded. Republican military authorities were instructed to introduce *fêtes décadaires* to the liberated lands, and the calendar and the new metric measurements were imposed by officials on Switzerland, Belgium, Italy, parts of Germany and the Netherlands. It was, historians suggest, received coolly. It seems likely that other than republican officials, eager local Jacobin sympathisers and the military, the practical impression made by the calendar beyond France was minimal at best. If the calendar travelled at all, it did so on the mastheads of pro-republican newspapers.[2] Military adventuring also brought France into contact with the Ottoman Empire and the Islamic Calendar; and, as heirs to the Enlightenment, there was a certain amount of interest in the lunar system used in the Ottoman Empire. During the expedition to Egypt, the French ordered the printing of an almanac in Cairo from the Year VIII, adjusted for the eastern longitude. Several concordances between the Islamic and republican calendars were published or submitted to the committee for public instruction in the 1790s and 1800s, a reminder of the links between politics, diplomacy and culture.

The calendar was also exported to the Americas and the Caribbean, where

1 Perovic, 'Turbulent time'.

2 Klein, *The science of measurement*, 125; Blanning, *The French Revolution in Germany*, 225.

it was employed by republican officials in the remaining French colonies. International treaties, such as the Louisiana Purchase, were conducted according to the republican calendar on the French side, and a puritan minister in Massachusetts reflected on its meaning, finding that it would help to obliterate such 'superstitions' as 'saints days, feasts, and fasts., etc., which make a great part of the Romish church' (the New England clergy remained a strong supporter of the Revolution, viewing it as a parallel to the United States' own quest for liberty).[3] The calendar was not restricted to diplomatic circles or officials, however, for some locals also attempted to impose the new order of time. In 1795 Victor Hugues, the Jacobin commissaire in Guadeloupe, struggled to encourage the freed slaves to work 'like true republicans'. In the spring of that year, he noted that workers took Sundays off to cultivate their own crops, while other workers took the *décadi*. Hugues, wishing to help them 'forget everything that reminds them of their time in servitude', decreed that all workers could take the *nonidi* as a day for 'personal affairs' and reiterated that the *décadi* had now become the uniform day of rest. Since the former slaves remained unpaid and traditionally were allowed to work their own plots on the day of rest, it seems likely that plantations found their labourers absent on both Sundays and *décadis*, reflecting the patterns of work in France suggested in the earlier chapter.[4]

Like metrification, which was abolished during the Restoration, the republican calendar had strong political associations that were unsuitable for France's enemies. Moreover in Europe after the Vienna Congress, as *La Breviaire du républicain, ou les loisirs d'un sans-culotte* implied, it had also become closely linked to anticlericalism. If anything, the calendar came to stand as a cipher for the failures of the Revolution. France's enemies had found it a fruitful source of satire and ridicule, which could be manipulated for local political ends.

Britain, as bastion of anti-Revolution and anti-republican sentiment, dubbed the months of the calendar with names such as 'Sneezy' and 'Wheezy'.[5] James Gillray's caricature *Shrine at St Ann's Hill* (1798) (*see* figure 10) mocked Charles James Fox, portraying the Whig politician in clerical garb, unshaven as usual, and praying before a republican parody of the Ten Commandments. In an echo of the requirement to remember the Sabbath

3 S. Stillman, *Thoughts on the French Revolution*, Boston 1795, 17–20, quoted in G. B. Nash, 'The American clergy and the French Revolution', *William and Mary Quarterly* 3rd ser. xxii (1965), 395–6.

4 L. Dubois, '"The price of liberty": Victor Hughes and the administration of freedom in Guadaloupe, 1794–1798, *William and Mary Quarterly* 3rd ser. lvi (1999), 384.

5 Doyle, *Oxford history*, 442.

Figure 10. James Gillray, 'Shrine at St Ann's Hill', hand-coloured etching, London 1798 (British Museum Satires 9217)

and keep it holy, the fourth republican commandment demands: 'Thou shalt work nine days and do whatever you want on the tenth'. The calendar was not always seen in such an insidious or comedic light. In the early days of the Revolution, before William Pitt's clampdown on republican or revolutionary dissent, a range of English equivalents to the calendar, such as the Patriot's Calendar, were produced. Print shops in the young American republic also produced their own versions.[6]

6 *Greenleaf's New-York, Connecticut, & New-Jersey almanack*, New York.

The calendar and memory

Within France itself, the calendar's influence continued beyond its abolition in 1806, retaining a symbolic power as an alternative to other conceptions of politics, history and time. Partly because of its rejection by the Napoleonic regime, it could be used as shorthand for the Jacobin Revolution in particular. For example, Vigny's *Les Consultations du Docteur Noir* (1832) used the calendar as a device to locate its characters in the Revolutionary moment, suggesting that a character bore the marks of both the royal almanac and the republican calendar.[7] In less specific ways, it is possible to argue that the experience of time – its measurement and commemoration – during the 1790s and early 1800s left important legacies for France in the years to come. The festive year had become secularised, and subsequent regimes recognised the importance of orchestrating and controlling national festivals, balancing the need to shore up political support without creating dangerous, revolutionary 'crowds'. Memorialisation, commemoration and the representation of the national past had become an important component of the state. Although Napoleon abolished the calendar, his desire to assert the regime's and his own place in history remained, and Bonapartism, as Sudhir Hazareesingh has argued, offered a means of marshalling true, popular festivals.[8] Napoleon also underscored the turn to what might be called the 'historical mode', which emphasised the linear temporality of the nation (and the powerful individual). The emperor himself was fond of historical literature and was acutely aware of the importance of history and his stature within it. His regime sought legitimisation in the past through the appropriation of such symbols as the bees of Charlemagne's crest on the emperor's cloak during his coronation at Notre Dame in 1804. He also instigated prizes for historical writing; commissioned paintings of his military triumphs in the historical mode; and, once ensconced on the rock of St Helena, set about enshrining his place in the pages of history through the production of memoirs and conversations with a stream of literate and well-connected visitors. The nineteenth century would witness the creation of the cult of Napoleon, a desire for past days of glory which Napoleon III could tap into following the financial crises of the July monarchy.[9]

7 A. Vigny, *Les Consultations du Docteur Noir: Stello ou les diables bleus*, Paris 1832, ch. xxiii.

8 Ozouf, *Festivals and the French Revolution*; S. Hazareesingh, *The Saint-Napoleon: celebrations of sovereignty in nineteenth-century France*, Cambridge, MA 2004, 224–5.

9 Hazareesingh, *Saint-Napoleon*.

The Bourbon Restoration responded to the republic's and the empire's concern with public ceremony, history and time, with its own message about the place of the French nation in time. The Restoration, which sought to 're-forge the links with the past', outlawed the forms and symbols of republican France, presenting France as a nation which had sinned, but had now been delivered 'from the grip of the Antichrist' and returned to its historical destiny as a Catholic, monarchical state. As Sheryl Kroen has shown in her analysis of the political and religious culture of the Restoration, the newly reinstated monarchy required the nation as a whole 'to accept responsibility for the sins of the Revolution, publicly confess, and erect symbols of repentance and renewed devotion to the true God and his representative on earth, the king'. Such expiation demanded 'remembering [*sic*] the sins of the past in detail' and missionaries across France reawakened revolutionary memories in attempts to resacralise former sites of liberty trees or guillotines. Fresh matrimonial rites replaced civic marriages, as in Salies-de-Béarn in Aquitaine, where 238 couples remarried in church in 1818. Outdoor public masses were explicitly compared to former republican festivals, which were described as 'odious ceremonies' by the bishop of Strasbourg in 1825 when he announced the arrival of a Catholic mission to the city. In response to the years of suppression, the Church vigorously reintroduced the festivals and ceremonies of the Catholic year.[10] Partly in response to the secularisation of the calendar, the Restoration Catholicised the yearly cycle, contributing to a public piety that was perhaps more intense than that of the eighteenth century. The judicial machinery of the state also began to defend Sunday observance. On 7 June 1814 the comte de Beugnot, Louis XVIII's chief of police, published an ordinance forbidding Sunday trading or working. Public amusements were forbidden before 5 p.m., and the vending of alcohol was restricted. A series of further police measures restricting dominical activities was codified in the law of 18 November 1814; a measure that was generally unwelcome to a population that was by now used to taking care of their business as they pleased. These regulations were applied with varying degrees of enthusiasm by departmental authorities throughout the nineteenth century until further reform in the 1880s.[11]

Nineteenth-century republicans also continued to deploy the suppressed calendar as a symbol of republicanism and the Revolution. For the left, Romme's system retained its republican associations, even finding them strengthened in the face of reaction. Key festivals, such as that commemo-

10 Kroen, *Politics and theater*, ch. iii.
11 Beck, *Histoire du dimanche*, 159–67.

rating the death of the king, were celebrated by republicans; such associations, as Kroen shows, continued throughout the nineteenth century and beyond.[12] The calendar retained its hold on the republican imagination, and on 2 May 1871 the council of the ill-fated Paris Commune restored both the committee of public safety and the republican calendar. Grenoble also copied the system.[13] Auguste Comte, the 'founder of the humanist religion', and his followers published a *Calendrier positiviste*, in which they named 1 January 1789 as 'le début de l'ère propre au Calendrier historique'. Comte proposed a year of thirteen months, each of twenty-eight days, named after men such as Moses, Homer, Cesar, Saint Paul, Gutenberg and Bichat; each month was also dedicated to a historic and improving theme, from ancient poetry to modern science. The final day of the year was dedicated to a festival of the dead, and in the leap year to 'saintes-femmes'.[14] Nor were the decimal reforms of the calendar forgotten in advanced circles. In 1908 a member of the Société de Géographie de Toulouse was pleased to report that thanks to the society's influence a Paris marine clockmaker had produced a number of decimal clocks to be sold in France and in Russia, Portugal and Romania. As a result of its industrial design, the piece, described as 'a fine decimal watch, with three hands', could be sold for fifty *francs*. The inventor of the clock, who praised the recent success of decimal weights and measures, considered that the decimal reform of time 'is also on the way'.[15]

The poetic and imaginative associations of the months of the republican year have also reverberated in western culture. The title of Zola's novel, *Germinal*, alluded to the birth of a proletarian, republican France that the story of the striking miners considered, while 'thermidor' became political shorthand for moderate reaction following a revolution. Radical political reformers and revolutionaries have also often followed the French example and torn up the existing calendar. During the Revolutions of 1848, posters and periodicals in Berlin proclaimed that it was 'Jahre 1' of liberty, and elsewhere in Europe communitarian and socialist groups experimented with their own calendars.[16] Within France, the primary legacy of the calendar can be

12 French anti-globalisation groups have used the calendar on the internet: www. decadi. com.

13 R. Christiansen, *Tales of the new Babylon: Paris, 1869–1875*, London 1994, 333. Official reports were dated from May by the republican calendar: *Pall Mall Gazette*, Feb. 1870.

14 A. Comte, *Calendrier positiviste*, Paris 1891.

15 Rey-Pailhade, 'Étude historique', 429–57.

16 F. Buchhändler, *General-Versammlung der fliegenden Buchhändler Berlins. Abgehalten am*, iii, Berlin [1848]; A. F. Karbe, *Karbe's Sendschreiben an seine Mitbürger*, [Berlin] [1848]. I am grateful to Susan Reed, curator of German and Incunabula at the British Library for this reference.

seen to be the close relationship between the nation and festivals. The Third Republic sought, with no little success, to perpetuate the memory of the Revolution and the republic through the creation of Bastille Day. In the second decade of the twentieth century, Bolshevik Russia swiftly disposed of the Orthodox Calendar, experimenting under Stalin with a number of 'rational' calendars based on groups of workers following different patterns of five-day weeks with a series of five festivals. Lenin's birthday became a national holiday. In September 1937 Stalin proposed international calendar reform at the League of Nations (without success).[17] The notion of a secular, national festival has since become commonplace in Europe and beyond, with liberation or foundation days regularly being commemorated across the globe.

More prosaically, the state also learned to intervene in the calendar, beginning to take steps along the path towards regularised work patterns, set holidays and the breaking of many of the links between the Christian and the secular year, a process begun during the *ancien régime* and intensified during the revolutionary and Napoleonic periods.[18] The high days of Assumption and Easter may have become public holidays, with protection for workers (if the holiday falls during the working week), but legal contracts no longer recognised such dates as markers for rent payments and so forth. As Meinzer has suggested, one of the republican calendar's most enduring legacies was the rationalised, state-controlled system of festivals as seen in the organisation of national holidays, such as the Napoleonic commemoration of the battle of Austerlitz, and limits on working time. The calendar also marked a conscious move to a secular, religiously neutral calendar, which despite its adoption by dechristianisers, was promoted as a tool of toleration that privileged neither Catholic nor Protestant.

Globalisation and the experience of colonialism has also reduced the number of timekeeping systems around the world, while advances in communications and travel technology have forced the world to find ways to orchestrate its temporal rhythms. Today, it seems improbable that the densely textured, ceremonially coloured and resolutely local calendar of the *ancien régime* could have survived into the modern world. Those living outside of France may miss the playful sincerity of 14 July or other, often local, celebrations that France seems still to view as a natural part of life; the importance

17 R. C. Williams, 'The Russian Revolution and the end of time, 1900–1940', *Jahrbücher für Geschichte Ost europas* n.s. xliii (1995), 364–401.

18 Ibid. The 'Year Zero' of Cambodia was more a matter of rhetoric and ideology than actual calendar reform, notes D. Chandler, *A history of Cambodia*, Boulder, Co 2008, 332 n. 14: 'The Khmer Rouge did not use 'year zero' or 'year one' in their propaganda, but they often spoke of Cambodian history having ended and begun on April 17 [1975]'.

placed on the summer holiday; the marches of 1 May; or the commercial quietness of a French town on a Sunday; but it is likely that the calendar is increasingly as much an invisible part of life there, denuded of much of its symbolism and sense of difference between the ordinary and the festive, as it is in the Anglo-Saxon world. The clinical precision of Romme's early plans for the new calendar, with the days of the week numbered from 'one' to 'ten', perhaps evoke this empty modernity, but, as this book has demonstrated, the reality is far richer, and far more varied, than that Enlightenment vision allowed.

APPENDICES

APPENDIX 1

Timeline of Key Events, 1788–1806

1788	
8 January	Sylvain Maréchal imprisoned for publishing his *Almanach des honnêtes gens.*
1789	Submission of the *cahiers des doléances* (books of grievances). Some included calls to reform weights and measures and reduce the surfeit of fêtes.
5 May	Opening of the *États-Généraux.*
17 June	Creation of the National Assembly.
20 June	Tennis Court Oath.
27 June	Commissioners appointed to reform weights and measures.
9 July	Creation of the Constituent Assembly.
14 July	Taking of the Bastille. Journals and pamphlets designate 1789 as the 'First Year of Liberty'.
1790	
8 May	Division of the arc of the meridian deemed to be the basis of measurement.
17 May	Lalande proposes changing the calendar.
12 July	Civil Constitution of the Clergy.
14 July	Fête de la Fédération; *Le Moniteur* masthead dated '1er jour de la deuxième année de la Liberté'.
1791	
21 June	Capture of the fleeing royal family at Varennes.
14 October	Decree creating the committee of public instruction
1792	
2 January	Legislative Assembly decrees that all public, judicial, civil and diplomatic records carry the inscription 'IV Year of the era of Liberty'. Each year of the era of liberty is deemed to begin on 1 January.

10 August	Storming of the Tuileries and convocation of National Convention.
21 August	*Le Moniteur* records the '4th year of liberty' and the '1st year of equality'.
2 September	Paris prison massacres.
21 September	France becomes a republic.
22 September	Convention declares that all public acts will be dated 'Year I of the Republic'.
21 December	Romme asked to bring the common and republican eras 'into accord'.

1793/ Year II

2 January	The Convention announces the beginning of the 2nd year of the Republic.
21 January	Execution of Louis XVI.
1 August	Decree on system of weights and measures.
20 September	Presentation of Romme's report on the new era. Discussion of the names of the months in the Convention.
5 October	Adoption of the new calendar.
7 October	Public servants to rest only on 10, 20 and 30 of each month.
18 October	Chénier, David, Fabre d'Églantine and Romme charged with providing new names for the months.
9 November (19 *brumaire*)	Decree combining all acts on the calendar into one. Fourth year of each *Franciade* to be called the 'année sextile'.
24 November (4 *frimaire*)	Single decree on the calendar, publication of instructions for the republican calendar.
7 December	Decree ordering the use of decimal money.

1794

18 February (30 *pluviôse*)	Publication of Romme's *Annuaire du cultivateur*.
8 June (20 *prairial*)	*Fête de l'Être suprême*.
27 July (9 *thermidor*)	Fall of Robespierre.

Year III

12 October (21 *vendémiaire*)	*Fête du calendrier* held in Arras.

1795

28 February (10 *ventôse*)	Abandonment of decimal time.
18 April (29 *germinal*)	Discussion of reform of leap years.
8 May (19 *floréal*)	Romme proposes modification of the calendar.
20 May (1 *prairial*)	Insurrection demanding bread and 1793 Constitution; a number of deputies, including Romme, support the rebels.
17 June (29 *prairial*)	Suicide of Romme following his arrest and imprisonment.
24 August (7 *fructidor*)	*Jours complémentaires* replace the *jours sans-culottes*.

Year IV

25 September (3 *vendémiaire*)	Adoption of the Constitution. Article 373 states that the '*èra franeaise* begins on 22 September 1792, the day of the foundation of the Republic'.

Year VI

1798

3 April (14 *germinal*)	Decree on the use of the republican calendar.
4 August (17 *thermidor*)	Decree on the observation of the *décadi*.
10 September (24 *fructidor*)	Law on the observance of the republican calendar.

Year VIII

1799

9 November (18 *brumaire*)	*Coup d'état*, bringing Napoleon Bonaparte into the government.

Year X

1802

8 April (18 *germinal*)	Re-establishment of Sunday as day of rest for officials.

Year XI

1803

1 January (11 *nivôse*)	*Le Moniteur* dates its issues by the Gregorian Calendar.

Year XIII

1805

9 September (22 *fructidor*)	Sénat approves Laplace's report on the suppression of the republican calendar.

1806

1 January	Reintroduction of the Gregorian Calendar.

APPENDIX 2

The Republican Calendar: a Glossary

(Autumn)

Vendémiaire	Vintage
Brumaire	Fog
Frimaire	Frost

(Winter)

Nivôse	Snow
Pluviôse	Rain
Ventôse	Wind

(Spring)

Germinal	Germination
Floréal	Flowering
Prairial	Meadow

(Summer)

Messidor	Harvest
Thermidor	Heat
Fructidor	Fruit

Divisions of time

Seconde décimale	100 seconds per *minute décimale*
Minute décimale	100 minutes per *heure décimale*
Heure décimale	10 hours per solar day
Décade	10 days per *décade*
Mois	3 *décades* per month
Sans-culottides/ Jours complémentaires	Final five (or six days) of the year earmarked for festivals
Année	12 months and 5 days per year
Franciade	4 years and 1 day

Franciade séculaire	100 *franciades simples*, less 3 days
Franciade millaire	10 *franciades séculaires*, less 1 day

Republican days of the *Décade*

Primidi, duodi, tridi, quartidi, quintidi, sextidi, septidi, octidi, nonidi, décadi

APPENDIX 3

Names of the Days of the Republican Year

	Autumn						**Winter**				
	Vendémiaire		**Brumaire**		**Frimaire**		**Nivôse**		**Pluviôse**		**Ventôse**
1	Raisin	1	Pomme	1	Raiponce	1	Tourbe	1	Lauréole	1	Tussilage
2	Safran	2	Célari	2	Turneps	2	Houille	2	Mousse	2	Cornouiller
3	Châtaigne	3	Poire	3	Chicorée	3	Bitume	3	Fragon	3	Violier
4	Colchique	4	Betterave	4	Nèfle	4	Soufre	4	Perce-neige	4	Troène
5	Cheval	5	Oie	5	Cochon	5	Chien	5	Taurreau	5	Bouc
6	Balsamine	6	Héliotrope	6	Mâche	6	Lave	6	Laurier-thym	6	Asaret
7	Carotte	7	Figue	7	Chou-fleur	7	Terre végétale	7	Amadouvier	7	Alaterne
8	Amaranthe	8	Scorsonère	8	Miel	8	Fumier	8	Mézéréon	8	Violette
9	Panais	9	Alisier	9	Genièvre	9	Salpêtre	9	Peuplier	9	Marceau
10	CUVE	10	CHARRUE	10	PIOCHE	10	FLÉAU	10	COIGNÉE	10	BÊCHE
1	Pomme-de-terre	1	Salsifis	1	Cire	1	Granit	1	Ellébore	1	Narcisse
2	Immortelle	2	Mâcre	2	Raifort	2	Argile	2	Brocoli	2	Orme
3	Potiron	3	Topinambour	3	Cèdre	3	Ardoise	3	Laurier	3	Fumeterre
4	Réséda	4	Endive	4	Sapin	4	Grès	4	Avelinier	4	Vélar
5	Âne	5	Dindon	5	Chevreuil	5	Lapin	5	Vache	5	Chèvre
6	Belle-de-nuit	6	Chervis	6	Ajonc	6	Silex	6	Buis	6	Epinards
7	Citrouille	7	Cresson	7	Cyprès	7	Marne	7	Lichen	7	Doronic
8	Sarrasin	8	Dentelaire	8	Lierre	8	Pierre à chaux	8	If	8	Mouron
9	Tournesol	9	Grenade	9	Sabine	9	Marbre	9	Pulmonaire	9	Cerfeuil
10	PRESSOIR	10	HERSE	10	HOYAU	10	VAN	10	SERPETTE	10	CORDEAU
1	Chanvre	1	Bacchante	1	Érable à sucre	1	Pierre à plâtre	1	Thlaspi	1	Mandragore
2	Pêche	2	Azerole	2	Bruyère	2	Sel	2	Thymelé	2	Persil
3	Navet	3	Garance	3	Roseau	3	Fer	3	Chiendent	3	Cochléaria
4	Amaryllis	4	Orange	4	Oseille	4	Cuivre	4	Trainasse	4	Pâquerette
5	Bœuf	5	Faisan	5	Grillon	5	Chat	5	Lièvre	5	Thon
6	Aubergine	6	Pistache	6	Pignon	6	Étain	6	Guède	6	Pissenlit
7	Piment	7	Macjonc	7	Liège	7	Plomb	7	Noisetier	7	Sylvie
8	Tomate	8	Coing	8	Truffe	8	Zinc	8	Cyclamen	8	Capillaire
9	Orge	9	Cormier	9	Olive	9	Mercure	9	Chélidoine	9	Frêne
10	TONNEAU	10	ROULEAU	10	PELLE	10	CRIBLE	10	TRAÎNEAU	10	PLANTOIR

	Spring **Germinal**		**Floréal**		**Prairial**		Summer **Messidor**		**Thermidor**		**Fructidor**
1	Primevère	1	Rose	1	Luzerne	1	Seigle	1	Épeautre	1	Prune
2	Platane	2	Chêne	2	Hémérocalle	2	Avoine	2	Bouillon-blanc	2	Millet
3	Asperge	3	Fougère	3	Trèfle	3	Oignon	3	Melon	3	Lycoperdon
4	Tulipe	4	Aubépine	4	Angélique	4	Véronique	4	Ivraie	4	Escourgeon
5	Poule	5	Rossignol	5	Canard	5	Mulet	5	Bélier	5	Saumon
6	Bette	6	Ancolie	6	Mélisse	6	Romarin	6	Prêle	6	Tubéreuse
7	Bouleau	7	Muguet	7	Fromental	7	Concombre	7	Armoise	7	Sucrion
8	Jonquille	8	Champignon	8	Martagon	8	Échalote	8	Carthame	8	Apocyn
9	Aulne	9	Hyacinthe	9	Serpolet	9	Absinthe	9	Mûre	9	Réglisse
10	COUVOIR	10	RÂTEAU	10	FAUX	10	FAUCILLE	10	ARROSOIR	10	ÉCHELLE
1	Pervenche	1	Rhubarbe	1	Fraise	1	Coriandre	1	Panic	1	Pastèque
2	Charme	2	Sainfoin	2	Bétoine	2	Artichaut	2	Salicorne	2	Fenouil
3	Morille	3	Bâton-d'or	3	Pois	3	Girofle	3	Abricot	3	Épine-vinette
4	Hêtre	4	Chamerisier	4	Acacia	4	Lavande	4	Basilic	4	Noix
5	Abeille	5	Ver à soie	5	Caille	5	Chamois	5	Brebis	5	Truite
6	Laitue	6	Consoude	6	Œillet	6	Tabac	6	Guimauve	6	Citron
7	Mélèze	7	Pimprenelle	7	Sureau	7	Groseille	7	Lin	7	Cardère
8	Ciguë	8	Corbeille-d'or	8	Pavot	8	Gesse	8	Amande	8	Nerprun
9	Radis	9	Arroche	9	Tilleul	9	Cerise	9	Gentiane	9	Tagette
10	RUCHE	10	SARCLOIR	10	FOURCHE	10	PARC	10	ÉCLUSE	10	HOTTE
1	Gainier	1	Statice	1	Barbeau	1	Menthe	1	Carline	1	Églantier
2	Romaine	2	Fritillaire	2	Camomille	2	Cumin	2	Câprier	2	Noissette
3	Marronnier	3	Bourrache	3	Chèvrefeuille	3	Haricot	3	Lentille	3	Houblon
4	Roquette	4	Valériane	4	Caille-lait	4	Orcanète	4	Aunée	4	Sorgho
5	Pigeon	5	Carpe	5	Tanche	5	Pintade	5	Loutre	5	Écrevisse
6	Lilas	6	Fusain	6	Jasmin	6	Sauge	6	Myrte	6	Bigarade
7	Anémone	7	Civette	7	Verveine	7	Ail	7	Colza	7	Verge-d'or
8	Pensée	8	Buglosse	8	Thym	8	Vesce	8	Lupin	8	Maïs
9	Myrtille	9	Sénevé	9	Pivoine	9	Blé	9	Coton	9	Marron
10	GREFFOIR	10	HOULETTE	10	CHARIOT	10	CHALÉMIE	10	MOULIN	10	PANIER

APPENDIX 4

Concordance for the Gregorian and Republican Calendars

Table 1
French republican calendar to the Gregorian Calendar

	Year I (1792–3)		Year VIII 1799–1800	
	Year II 1793–4		Year IX 1800–1	
	Year III 1794–5*		Year X 1801–2	
	Year V 1796–7		Year XI 1802–3*	
	Year VI 1797–8		Year XIII 1804–5	
	Year VII 1798–9*	Year IV 1795–6	Year XIV 1805	Year XII 1803–4
1 *vendémiaire*	22 Sept.	23 Sept.	23 Sept.	24 Sept.
1 *brumaire*	22 Oct.	23 Oct.	23 Oct.	24 Oct.
1 *frimaire*	21 Nov.	22 Nov.	22 Nov.	23 Nov.
1 *nivôse*	21 Dec.	22 Dec.	22 Dec.	23 Dec.
1 *pluviôse*	20 Jan.	21 Jan.	21 Jan.	22 Jan.
1 *ventôse*	19 Feb.	20 Feb.	20 Feb.	21 Feb.
1 *germinal*	21 Mar.	21 Mar.	22 Mar.	22 Mar.
1 *floréal*	20 Apr.	20 Apr.	21 Apr.	21 Apr.
1 *prairial*	20 May	20 May	21 May	21 May
1 *messidor*	19 June	19 June	20 June	20 June
1 *thermidor*	19 July	19 July	20 July	20 July
1 *fructidor*	18 Aug.	18 Aug.	19 Aug.	19 Aug.
jour de la Vertu	17 Sept.	17 Sept.	18 Sept.	19 Sept.
jour du Génie	18 Sept.	18 Sept.	19 Sept.	20 Sept.
jour du Travail	19 Sept.	19 Sept.	20 Sept.	21 Sept.
jour de l'Opinion	20 Sept.	20 Sept.	21 Sept.	22 Sept.
jour des Récompenses	21 Sept.	21 Sept.	22 Sept.	23 Sept.
* In leap year also *jour de la Révolution*	22 Sept. 1795, 1799		23 Sept. 1803	

Table 2
Gregorian Calendar to the French republican calendar

	1793 Years I–II			1800 Years VIII–IX		
	1794 Years II–III			1801 Years IX–X		
	1797 Years V–VI	1795 Years III–IV		1802 Years X–XI		
	1798 Years VI–VII	1799 Years VII–VIII	1796 Years IV–V	1805 Years XIII–XIV	1803 Years XI–XII	1804 Years XII–XIII
1 Jan.	12 *nivôse*	12 *nivôse*	11 *nivôse*	11 *nivôse*	11 *nivôse*	10 *nivôse*
1 Feb.	13 *pluviôsc*	13 *pluviôse*	12 *pluviôse*	12 *pluviôse*	12 *pluviôse*	11 *pluviôse*
1 Mar.	11 *ventôse*	11 *ventôse*	11 *ventôse*	10 *ventôse*	10 *ventôse*	10 *ventôse*
1 Apr.	12 *germinal*	12 *germinal*	12 *germinal*	11 *germinal*	11 *germinal*	11 *germinal*
1 May	12 *floréal*	12 *floréal*	12 *floréal*	11 *floréal*	11 *floréal*	11 *floréal*
1 June	13 *prairial*	13 *prairial*	13 *prairial*	12 *prairial*	12 *prairial*	12 *prairial*
1 July	13 *messidor*	13 *messidor*	13 *messidor*	12 *messidor*	12 *messidor*	12 *messidor*
1 Aug.	14 *thermidor*	14 *thermidor*	14 *thermidor*	13 *thermidor*	13 *thermidor*	13 *thermidor*
1 Sept.	15 *fructidor*	15 *fructidor*	15 *fructidor*	14 *fructidor*	14 *fructidor*	14 *fructidor*
1 Oct.	10 *vendémiaire*	9 *vendémiaire*	10 *vendémiaire*	9 *vendémiaire*	8 *vendémiaire*	9 *vendémiaire*
1 Nov.	11 *brumaire*	10 *brumaire*	11 *brumaire*	10 *brumaire*	9 *brumaire*	10 *brumaire*
1 Dec.	11 *frimaire*	10 *frimaire*	11 *frimaire*	10 *frimaire*	9 *frimaire*	10 *frimaire*

Source: B. Blackburn and L. Holford-Strevens, *The Oxford companion to the year*, Oxford 1999, 850–1. Reproduced by kind permission of Oxford University Press.

Bibliography

Unpublished primary sources

Amiens, Archives départmentales Somme
L 421

Avignon, Archives départmentales Vaucluse
18 L 24

Besaneon, Archives départmentales Doubs
L 246, L 316, L368, L 369, L 377, L 380

Besaneon, Archives municipales
DD 176
HH1
I1 31, XI, I1 31
I3 1
M1 1
U 7592

Dijon, Archives départmentales Côte-d'Or
L 472, L 476, L 1660, L 1701

Dijon, Archives municipales
D 29

Lille, Archives départmentales Nord
L 1265, L 1267, L 5045, L 5046

Milan, Museo del risorgimento
A2 Romme, 1794

Nantes, Archives départmentales Loire Atlantique
L 99, L 152, L 153, L 369, L 2113*, L 2430

Nantes, Archives municipales
I2 8 9

Paris, Archive de la Préfecture de Police de Paris
AA 202, AA 236

Paris, Archives Nationales
AD VIII 18 5, AD VIII 18 15
AF III 47
C 445, C 561
$D^{III}37$
F 1331, F 1355
$F^{1c}184$
F^{1c} III Doubs 1, F^{1c} III Doubs 11, F^{1c} III Drome 11, F^{1c} III Eure 12, F^{1c} III Gironde 8, F^{1c} III Gironde 9, F^{1c} III Haute-Garonne 12, F^{1c} III Isère 9, F^{1c} III Loire-Inférieure 9, F^{1c} III Loire-Inférieure 10, F^{1c} III Pud-de-Dôme 9, F^{1c} III Vaucluse 7, F^{1c} III Vaucluse 8, F^{1c} III Vaucluse 9
$F^{4}2334$
$F^{7}382$
F^{12} 1274, $F^{12}1326$, $F^{12}2403$, $F^{12}2406$, $F^{12}2434$
F^{17} 1008 D, $F^{17}1135$, $F^{17}1338$, $F^{17}1355$, $F^{17}1454$, $F^{17}3702$
$O^{1}3510$
Y 9498, Y 9499, Y 12503

Paris, Bibliothèque nationale
NAF 4789

Paris, Bibliothèque de l'Observatoire de Paris
MS B 57

Microfilm
French Revolution Research Collection, Oxford 1990–

Printed primary sources

Newspapers
L'Ancien Moniteur
Annales de religion
Courrier républicain
Courrier des spectacles ou Journal des théâtres
Journal du matin des Amis de Liberté et de l'Égalité
Journal révolutionnaire de Toulouse ou le surveillant du Midi
Le Moniteur
Pall Mall Gazette
Paris periscope
The Times

The True Briton
La Vedette ou Journal du département du Doubs

Contemporary books and articles

Académie des Sciences, *Sur l'Uniformité et le systême général des poids et measures: rapport et projet de décret présentées à la Convention nationale par le citoyen Arbogast et rapport fait a l'académie des sciences par les citoyens Borda, Lagrange et Monge*, Paris 1793
[A. Alison], 'Fall of the throne of the barricades', *Blackwood's Edinburgh Magazine* lxiii (1848), 393–418
Almanach de gotha: contenant diverses connaissances curieuses et utiles pour l'année, Gotha 1763
Almanach nécessaire, ou porte-feuille de tous les jours, utile à toutes personnes, Paris 1789
Almanach des prisons, Paris Year III
Annales de la religion; ou, mémoires pour servir a l'histoire du dix-huitième (et XIXe) siècle: par une société d'amis de la religion et de la patrie, Paris 1795–1803
Annuaire ou calendrier pour la seconde année de la République Francaise, Paris 1793
Anon., *Il Étoit temps ou la semaine aux événemens*, [Paris] [1789]
Arrêté du Directoire Exécutif qui prescrit des mesures pour la stricte exécution du calendrier républicain, 14 germinal VI (3 Apr. 1798).
Barré, M., F. P. A. Leger and J. R. Rosières, *L'Heureuse Décade, divertissement patriotique, en un acte et en vaudevilles, des citoyens barré, léger et rosières*, Paris 1793, repr. in *Répertoire du théâtre républicain: ou, recueil de pièces imprimées avant, pendant et après la république francaise*, vii, Geneva 1986
Barthelemy, J.-J., *Voyage du jeune Anacharsis en Grèce*, Paris 1788
[A. Beers and J. Reid], *The New-York Connecticut, & New Jersey almanack, or diary, for the year of our Lord, 1799: ... Fitted for the meridian and horizon of New-York, but will serve without essential variation, for Connecticut and New Jersey. ... To which is added, with directions the new French calendar*, New York [1798]
Bellement, A.-P., *La Seconde Décade, ou le double mariage, opéra patriotique en 2 actes et en vaudeville, suite de l''Heureuse Décade'*, Paris 4 nivôse II [1793]
Les Bordels de Paris, avec les noms, demeures et prix, Paris [1790]
Buchhändler, F., *General-Versammlung der fliegenden Buchhändler Berlins. Abgehalten am*, iii, Berlin [1848]
Bugge, T., *Travels in the French republic: containing a circumstantial view of the present state of learning, the arts, learned societies, manners &c. in that country*, London 1801
Calès, J.-M., *Proclamation du représentant du peuple J. Marie Calès, délègue par la Convention nationale dans le département du Doubs, pour y protéger l'établissement de la manufacture d'horlogerie, aux citoyens de la commune de Besancon*, Besancon Year III
Carlyle, T., 'Terror', in *The French Revolution*, iii, London 1837

Catéchisme historique et révolutionnaire, contenant un recueil exact, authentique, chronologique et impartial des événements remarquables, ainsi que des actions d'éclat, traits de courage, d'héroïsme et autres qui sont arrivés, jour par jour, [n.p.] [n.d.]

Comte, A., *Calendrier positiviste*, Paris 1891

Condorcet, M. J. A. N. Caritat, marquis de, *Esquisse d'un tableau historique des progrès de l'esprit humain, suivie de réflexions sur l'esclavage des nègres, par Condorcet*, Paris 1822

Cornet, M. A. de, *Opinion sur la résolution relative aux fêtes décadaires*, Paris Year VI

Correspondence de Napoleon Ier, Paris 1858

Crèvecoeur, J. H. St John de, *Letters from an American farmer: describing certain provincial situations, manners, and customs*, London 1782

—— *Lettres d'un cultivateur américain*, Paris 1784

Decadario per l'anno VI. della repubblica Francese, che comincia ... 21 Settembre 1797 col Colendario Romano, Padua 1797

Dickinson, J., *Letters from a farmer in Pennsylvania: to the inhabitants of the British colonies*, Philadelphia 1774

Diderot, D., *Encyclopédie, ou, dictionnaire raisonné des sciences, des arts et des métiers, par une société des gens de lettres: mis en ordre & publié par* M. *Diderot ... & quant à la partie mathématique par* M. d'Alembert, Geneva 1772

Dubourg, B., *Chronographie, ou description des tems; contenant toute la suite des souverains de l'univers, & des principaux événemens de chaque siècle, depuis la création du monde jusqu'à présent en trente-cinq planches gravées en taille-douce, & réunies en une machine d'un usage facile & commode*, Paris 1753

Duchesne, J., *Catalogue des tableaux, dessins, estampes, livres, médailles, coquilles et curiosités; planches gravées, curiosités, ouvrages à figures, livres sur les arts, littérature: du cabinet du feu m. Louis Lafitte, peintre, catalogue des tableaux*, Paris 1828

Examen impartial de cette question: comment et dans quelle forme doit-on procéder aux délibérations de l'assemblée prochaine des Etats-généraux?, Besaneon 1789

Fabre d'Églantine, P. F., *Rapport fait à la Convention nationale dans la séance du 3 du second mois de la seconde année de la République Franeaise, au nom de la commission chargée de la confection du calendrier, par Ph. France. Na. Fabre d'Églantine*, Paris Year II

Faure, P. J. D. G., *Sur les Fêtes décadaires*, Paris Year III

Flavigny, J.-B., *Instructions pastorales sur l'accord de la religion catholique avec le gouvernement républicain, et sur la conciliation des devoirs essentiels du culte chrétien avec les lois de l'état en France*, Paris 1799

Greenleaf's New-York, Connecticut, & New-Jersey almanack, New York 1790–1801

Grégoire, H., *Mémoires de Grégoire, ancien évêque de Blois, député à l'Assemblée Constituante et à la Convention Nationale, sénateur, membre de l'Institut*, Paris 1808

Guesdon, F., *Opinion sur le second projet de résolution présenté par Bonnaire, le 19 messidor, en ce qui concerne la célébration des mariages au chef-lieu de canton, les seuls jours de décadi*, Paris Year VI

Heurtault-Lamerville, J. M., *Opinion de Heurtault-Lamerville, député du Cher & membre de la commission d'instruction publique, sur les fêtes décadaires: séance du 28 messidor an 6*, Paris 1798 (AN, AD VIII 18 15)

Karbe, A. F., *Karbe's Sendschreiben an seine Mitbürger*, [Berlin] [1848]

Keysler, J. G., *Travels through Germany, Bohemia, Hungary, Switzerland, and Lorrain*, London 1756

Lanjuinais, J. D., *Sur l'introduction du calendrier des tyrans*, Paris Year III

Leclerc, J. B., *Rapport sur les institutions relatives à l'état civil des citoyens*, Paris Year VI

Lemercier, E., *Opinion sur la résolution du 6 thermidor, relative aux fêtes décadaires*, Paris Year VI

—— *Rapport fait par Lemercier, Député de la Charente-Inférieure au nom d'une commission spéciale, composée des représentants du peuple Bordas, Missonet & Lemercier, sur la résolution du premier nivôse an 6, relative aux citoyens Mégevand & Trot, chefs de la manufacture nationale d'horlogerie de Besançon. Séance du 14 messidor an 6*, Paris Year VI (AN, F^{12}1326)

*Lettre de l'administrateur du diocèse de *** aux prêtres de ce diocèse*, Anvers 1799

Linné, C. von and A. M. Berger, *Calendarium florae*, Upsala 1756

Marcoz, J.-B.-P., *Objet et ordre des fêtes décadaires de la république française*, Paris Year III

Mare, N. de la, *Traité de la police, où l'on trouvera l'histoire de son établissement, les fonctions et les prérogatives de ses magistrats, toutes les loix et tous les réglemens qui la concernent: on y a joint une description historique et topographique de Paris … avec un recueil de tous les statuts et règlemens des six corps des marchands et de toutes les communautez des arts et métiers: par M. Delamare*, i, Paris 1729

Maréchal, S., *Almanach des honnêtes-gens* [Paris 1788]

Les Martyrs de prairial: textes et documents inédits, ed. F. Brunel and S. Goujon, Geneva 1992

La Meilleure Manier de délibérer et de voter dans une grande assemblée, n.p. 1788

Mémoires de Barras: membre du Directoire, ed. G. Drury, Paris 1895

Ménétra, J.- L., *Journal de ma vie: Jacques-Louis Ménétra, compagnon vitrier au 18e siècle*, ed. Daniel Roche, Paris 1982

Mercier, L.-S., *Le Nouveau Paris*, Paris 1798–9

—— *Le Tableau de Paris*, ed. J. Kaplow, Paris 1979

Montesquieu, C. de Secondat, *De l'Esprit des loix*, i, London 1757

Moreau de Jonnès, A., *Adventures in the Revolution and under the Consulate*, London 1929

Moreau de l'Yonne, A., *Opinion sur la résolution relative aux fêtes décadaires*, Paris 1798

Papers and diaries of a York family, 1764–1839, ed. E. Gray, London 1927

Periaux, P., *Tableau du concordance entre les divisions décimales et les divisions anciennes*, [Paris] 1794

—— *Concordance des calendriers républicain et grégorien, depuis 1793 jusques et compris l'an XIV*, Paris 1805

Prieur, C.-A., *Mémoire sur la nécessité et les moyens de rendre uniformes, dans le royaume, toutes les mesures d'étendue et de pesanteur [Texte imprimé] .... d'en régler tous les multiples et les subdivisions suivant l'ordre décuplé ... par M. Prieur ci-devant Du Vernois*, Dijon 1790

Procès-verbaux du comité d'instruction publique de la Convention nationale, ed. M. J. Guillaume, Paris 1891–1957

Programmes des professeurs de l'école centrale du département de l'Isère pour le cours de l'an VII, Grenoble Year VII (AN, F^{17}1338)

Rapport fait par Bonnaire du Cher, au nom des commissions d'instruction publique & des institutions républicaines, réunies sur les fêtes décadaires: séance du 19 messidor an 6, Paris 1798 (AN, ADVIII 18 5)

Rapport sur les questions relatives au nouveau système horaire, Paris Year IV (AN, F^{17}1135 85).

Réimpression de l'ancien Moniteur; seule histoire authentique et inaltérée de la Révolution Franeaise, depuis la réunion des E´tats-Généraux jusqu'au Consulat (Mai 1789–Novembre 1799), xix, Paris 1850

Relation de tout ce qui s'est passé en Provence, n.p. 1789

Romme, C., *Rapport sur l'ère de la République*, Paris 1793

—— *Annuaire du cultivateur, pour la 3e année de la république*, Paris [1794]

Rousseau, J. J., *Emilius and Sophia, or a new system of education*, ii, London 1783

Saint-Aubin, C., *Réflexions sur le nouveau calendrier et sur la nécessité d'élaguer au moins les décadis de cet avorton du système décimal; avec un mot en passant sur les effets du fanatisme politique comparés à ceux du fanatisme religieux*, [n.p.] 1795

Sérieys, A., *Introduction aux décades républicaines, ou à l'histoire abrégée de la république franeaise, etc.*, Paris Year II (1794)

Stendhal [M.-H. Beyle], *Oeuvres complètes de Stendhal*, ii, Paris 1923

Tench, W., *Letters written in France to a friend in London, between the month of November 1794 and the month of May 1795*, London 1796

Terral, J., *Réflexions sur les fêtes décadaires*, Paris 1795

Thiery, L. V., *Almanach du voyageur à Paris*, Paris 1789

Van Troil, U., 'Letters of Iceland', in J. Pinkerton (ed.), *Voyages and travels*, i, London 1808

Vigny, A. de, *Les Consultations du Docteur Noir: Stello ou les diables bleus*, Paris 1832

Young, A., *Travels in France during the years 1787, 1788, 1789*, Gloucester MA 1976

Secondary sources

Adam, B., *Timewatch: the social analysis of time*, Cambridge 1995

Agulhon, M., *Marianne au combat: l'imagerie et la symbolique républicaines de 1789 à 1880*, Paris 1979

Alder, K., *Engineering the revolution: arms and enlightenment in France, 1763–1815*, Princeton 1997

—— *The measure of all things: the seven-year odyssey and hidden error that transformed the world*, New York 2002

Anderson, B. R., *Imagined communities: reflections on the origin and spread of nationalism*, London 1983

Andress, D., *Massacre at the Champ de Mars: popular dissent and political culture in the French Revolution*, Woodbridge 2000

—— *The Terror: civil war in the French Revolution*, London 2005

Andrews, G. G., 'Making the Revolutionary Calendar', *AHR* xxxvi (1931), 515–32

Aston, N., *Religion and revolution in France, 1780–1804*, Washington, DC 2000

Aulard, F. A., *The French Revolution: a political history, 1789–1804*, trans. B. Miall, iii, London 1910

Baczko, B., *Lumières de l'utopie*, Paris 1978

—— 'Le Calendrier républicain', in Pierre Nora (ed.), *Les Lieux de mémoire*, Paris 1984, 37–83

—— *Ending the Terror: the French Revolution after Robespierre*, Cambridge 1994

Baker, K. M. (ed.), *Condorcet: from natural philosophy to social mathematics*, Chicago–London 1975

—— *The French Revolution and the creation of modern political culture*, i, Oxford 1987

—— *Inventing the Revolution*, Cambridge 1990

Barnes, B., 'Between the real and the reified: the sociology of time', in S. Loyal and S. Quilley (eds), *The sociology of Norbert Elias*, Cambridge 2004, 59–74

Baulant, M., 'Salaire des ouvriers du bailment à Paris de 1400 à 1726', *Annales: Economies, Sociétés, Civilizations* xxvi (1971), 463–83

Beck, R., *Histoire du dimanche: de 1700 à nos jours*, Paris 1997

Bell, D. A., *The cult of the nation in France: inventing nationalism, 1680–1800*, Cambridge, MA 2001

Benjamin, W., 'On the concept of history', http://www.marxists.org/reference/archive/benjamin/1940/history.htm

Bensimon, F., 'L'Écho de la Révolution Franeaise dans la Grande-Bretagne du XIXe siècle, 1815–1870', *AHRF* cccxlii (2005), 211–37

Bianchi, S., 'La Bataille du calendrier ou le décadi contre le dimanche: nouvelles approches pour la reception du calendrier republicain en milieu rural', *AHRF* cccxii (1998), 245–64

Blackburn B. and L. Holford-Strevens, *The Oxford companion to the year*, Oxford 1999

Blanning, T. C. W., *The French Revolution in Germany: occupation and resistance in the Rhineland, 1792–1802*, Oxford 1983

Bouchard, G., *Un Organisateur de la victoire: Prieur de la Côte-d'Or: membre du comité de salut public*, Paris 1946

Bourdin, P., *Des Lieux, des mots, les révolutionnaires: Le Puy-de-Dôme entre 1789 et 1799*, Clermont Ferrand 1995

Brelot, C., *Besancon révolutionnaire*, Besancon 1966

Brouillard, J., 'Un Ballet original', *Revue historique de Bordeaux et du département de la Gironde* v (1912), 277–8

Burrow, J., *A history of histories: epics, chronicles, romances & inquiries from Herodotus & Thucydides to the twentieth century*, London 2009

Burton, J. K., *Napoleon and Clio: historical writing, teaching, and thinking during the First Empire*, Durham 1979

Byrnes, J. F., *Catholic and French forever: religious and national identity in modern France*, University Park, PA 2005

Cardinal, C. (ed.), *La Révolution dans la mesure du temps: calendrier républicain heure décimale, 1793–1805*, La Chaux-de-Fonds 1989

Certeau, M. de, *The practice of everyday life*, Berkeley 1984

Chandler, D., *A history of Cambodia*, Boulder, CO 2008

Christiansen, R., *Tales of the new Babylon: Paris, 1869–1875*, London 1994

Clémendot, P., *Le Départment de la Meurthe à l'époque du Directoire*, Raon- l'Etape 1966

Cobb, R., 'Thermidor or the retreat from fantasy,' *Encounter* lvii (1981), 30–42

Cook, M., 'Politics in the fiction of the French Revolution, 1789–1794', *Studies on Voltaire and the Eighteenth Century* cci (1982), 233–340

—— *Fictional France: social reality in the French novel, 1775–1800*, Oxford 1993

Corbin, A., *Le Temps, le désir et l'horreur: essais sur le dix-neuvième siècle*, Paris 1990

—— *Village bells: sound and meaning in the 19th-century French countryside*, New York 1998

Crawley, C. W. (ed.), *New Cambridge modern history*, ix, Cambridge 1965

Crosby, A. W., *The measure of reality: quantification and western society, 1250–1600*, Cambridge 1997

Crow, T. E., *Emulation: making artists for revolutionary France*, New Haven–London 1994

Crowdy, T., *French Revolutionary infantry, 1789–1802*, Oxford 2004

Crowston, C. H., *Fabricating women: the seamstresses of Old Regime France, 1675–1791*, Durham, NC 2001

Darnton, R., 'What was revolutionary about the French Revolution?', *New York Review of Books*, 19 Jan. 1989, 3–10

Daumas, P., 'Prénoms et révolution, 1775–1825: propositions pour une nouvelle approche méthodologique', *Revue d'histoire moderne et contem- poraine* xliv (1997), 109–32

Desan, S., *Reclaiming the sacred: lay religion and popular politics in revolutionary France*, Ithaca 1990

Dohrn-van Rossum, G., *History of the hour: clocks and modern temporal orders*, Chicago 1996

Dommanget, M., 'Sylvain Maréchal, précurseur du calendrier révolutionnaire', *International Review for Social History* iii (1938), 301–34

Doyle, W., *The Oxford history of the French Revolution*, Oxford 1989

Dubois, L., '"The price of liberty": Victor Hugues and the administration of

freedom in Guadaloupe, 1794–1798', *William and Mary Quarterly* 3rd ser. lvi (1999), 363–92
Durand, E., *Mirage in the east; a history of the French image of American society to 1815*, Princeton, NJ 1957
Elias, N., *Time: an essay*, Oxford 1992
Ellis, H. A., *Boulainvilliers and the French monarchy: aristocratic politics in early eighteenth-century France*, Ithaca 1988
Emmet, K., *A cultural history of the French Revolution*, New Haven, CN 1989
Epstein, S. A., 'Business cycles and the sense of time in medieval Genoa', *Business History Review* lxii (1988), 238–60
Febvre, L., 'Travail: évolution d'un mot et d'une idée', *Journal de psychologie normale et pathologique* xli (1948), 19–28
Fehér, F. (ed.), *The French Revolution and the birth of modernity*, Berkeley 1990
Fontaine, L., *History of pedlars in Europe*, Oxford 1996
Forrest, A. I., *Society and politics in revolutionary Bordeaux*, Oxford 1975
—— *Paris, the provinces and the French Revolution*, London–New York 2004
Foucault, M., *Discipline and punish: the birth of the prison*, London 1977
Fox-Genovese, E., *The origins of Physiocracy: economic revolution and social order in eighteenth-century France*, Ithaca 1976
Frand, P., 'Le Calendrier républicain dans les annuaires de l'an IV a l'an XIIII', *AHRF* cciic (1994), 671–86
Fraser, J. T., *Time as conflict: a scientific and humanistic study*, Basle–Stuttgart 1978
Friguglietti, J., 'Gilbert Romme and the making of the French republican calendar', in D. G. Troyansky, A. Cismaru and N. Andrews, Jr (eds), *The French Revolution in culture and society*, Westport, CN 1991, 13–22
Froeschle, M., 'A Propos du calendrier républicain: Romme et l'astronomie', *AHRF* ccciv (1996), 303–25
Furet, F., *Revolutionary France, 1770–1880*, Oxford 1992
—— and M. Ozouf (eds), *A critical dictionary of the French Revolution*, Cambridge, MA 1989
Garrioch, D., *Neighbourhood and community in Paris, 1740–1790*, Cambridge 1986
—— *The making of revolutionary Paris*, Berkeley 2002
Garrone, A. G., *Gilbert Romme: histoire d'un révolutionnaire, 1750–1795*, Paris 1971
Gilbert, F. (ed.), *The progress of society in Europe: a historical outline from the subversion of the Roman Empire to the beginning of the sixteenth century*, Chicago–London 1972
Gillispie, C. C., *Pierre-Simon Laplace, 1749–1827: a life in exact science*, Princeton, NJ 1997
Glennie, P. and N. Thrift, 'Reworking E. P. Thompson's "time, work- discipline and industrial capitalism"', *Time & Society* v (1996), 275–99
Godechot, J., 'Les Papiers du Romme', *AHRF* xxvi (1954), 63–73
Gordon, D., *Citizens without sovereignty: equality and sociability in French thought, 1670–1789*, Princeton 1994
Gough, H., *The Terror in the French Revolution*, London 1988

Griffiths, A., *Prints for books: book illustration in France, 1760–1800*, London 2004
Gross, D., 'Temporality and the modern state', *Theory and Society* xiv (1985), 53–82
Gross, J.-P., *Fair shares for all: Jacobin egalitarianism in practice*, Cambridge 1997
Gutwirth, M., *The twilight of the goddesses: women and representation in the French Revolutionary era*, New Brunswick, NJ 1992
Hahn, R., *Pierre Simon Laplace, 1749–1827: a determined scientist*, Cambridge, MA 2005
Halbwachs, M., *On collective memory*, trans. L. A. Coser, Chicago 1992
Hampson, N., 'Les Ouvriers des arsenaux de la marine au cours de la Révolution Franeaise, 1789–1794', *Revue d'histoire économique et sociale* xxxix (1961), 269–329, 442–73
—— *A social history of the French Revolution*, London 1963
—— *The enlightenment*, London 1968
Hanson, P. R., *The Jacobin republic under fire: the federalist revolt in the French Revolution*, University Park, PA 2003
Hartmann, A.-L., 'Foires et marchés en Maine-et-Loire, 1789–1815', *Annales de Bretagne et des Pays d'Ouest* ci (1994), 61–79
Hauser, H., *Travailleurs et marchands dans l'ancienne France*, Paris 1929
Hazareesingh, S., *The Saint-Napoleon: celebrations of sovereignty in nineteenth-century France*, Cambridge, MA 2004
Higonnet, P., 'Reviewed works: Der Französische Revolutionskalender, 1792–1805: Planung, durchführung und Scheitern einer politischen Zeitrechnung, by Michael Meinzer', *AHR* xcviii (1993), 1265–6
—— *Goodness beyond virtue: Jacobins during the French Revolution*, Cambridge 1998
Hill, C., *Puritanism and revolution: studies in interpretation of the English Revolution of the 17th century*, London 1964
Hunt, L. A., *Politics, culture, and class in the French Revolution*, Berkeley 1984
—— *The family romance of the French Revolution*, London 1992
—— 'The world we have gained: the future of the French Revolution: presidential address to the American Historical Association, 2003', *AHR* cviii (2003), 1–19
Ingold, T., 'The temporality of the landscape', *World Archaeology* xxv (1993), 152–72
Jacob, A., *Le Travail, reflet des cultures du sauvage indolent au travailleur productif*, Paris 1994
Jeweler's Circular and Horological Review xxxvi (1898)
Jones, C., *The Longman companion to the French Revolution*, Harlow 1988
—— 'The great chain of buying: medical advertisement, the bourgeois public sphere, and the origins of the French Revolution', *AHR* ci (1996), 13–40
—— *The great nation: France from Louis XV to Napoleon*, London 2003
—— *Paris: biography of a city*, London 2004
Jones, P. M., *The peasantry in the French Revolution*, Cambridge 1988

Jullien, M.-A., *Essai sur l'emploi du temps, ou méthode qui a pour objet de bien régler l'emploi du temps*, Paris 1810
Kelsey, S., *Inventing a republic: the political culture of the English Commonwealth, 1649–1653*, Manchester 1997
Kennedy, E., *A cultural history of the French Revolution*, New Haven 1989
Kingston, R., 'The bricks and mortar of revolutionary administration', *FH* xx (2006), 405–23
Klein, H. A., *The science of measurement: a historical survey*, New York 1988
Konvitz, J. W., *The urban millennium: the city-building process from the early Middle Ages to the present*, Carbondale–Edwardsville 1985
Kroen, S., *Politics and theater: the crisis of legitimacy in restoration France, 1815–1830*, Berkeley 2000
Kula, W., *Measures and men*, Princeton, NJ 1986
Labrosse, C., 'L'Apport de l'étude des journaux de 1789 a la réflexion sur l'événement', in R. C. Grenoble (ed.), *Aux Origines provinciales de la évolution*, Grenoble 1990, 255–66
—— 'Le Temps immédiat dans la presse parisienne de 1789', in *L'Espace et le temps reconstruits: la Révolution Franeaise, une révolution des mentalités et des cultures?*, Aix-en-Provence 1990, 110–20
Labroue, H., *L'Esprit public en Dordogne pendant la révolution*, Paris 1911
Lamaison, P., 'Des Foires et des marchés en Haute-Lozère', *Études rurales* lxxvii–lxxx (1980), 199–300
Landes, D. S., *Revolution in time: clocks and the making of the modern world*, Cambridge, MA 1983
Le Goff, J., *Time, work and culture in the Middle Ages*, Chicago 1980
Lefebvre, G., *La Revolution Franeiase*, Paris 1951
Leith, J. A, *The idea of art as propaganda in France, 1750–1799: a study in the history of ideas*, [Toronto] 1965
—— *Space and revolution: projects for monuments, squares, and public buildings in France, 1789–1799*, Montreal–London 1991
Letzter, J. and R. Adelson, *Women writing opera: creativity and controversy in the French Revolution*, Berkeley 2001
Lick, R., 'Les Intérieurs domestiques dans la seconde moitié du XVIIIe siècle, d'après les inventaires après décès de Coutances', *Annales de Normandie* iv (1970), 293–316
Livesey, J., 'Agrarian ideology and commercial republicanism in the French Revolution', *Past & Present* clvii (1997), 94–121
—— *Making democracy in the French Revolution*, Cambridge, MA 2001
Lucas, C., *The structure of the Terror: the example of Javogues and the Loire*, Oxford 1973
Lyons, M., *France under the Directory*, Cambridge 1975
Macey, S. L., *Clocks and the cosmos: time in western life and thought*, Hamden, CN 1980
Maiello, F., *Storia del calendario: la misurazione del tempo, 1450–1800*, Turin 1994
Mainz, V., *L'Image du travail et la Révolution Franeaise*, Vizille 1999

Margadant, T. W., *Urban rivalries in the French Revolution*, Princeton 1992
Margairaz, D., *Foires et marchés dans la France préindustrielle*, Paris 1988
Margerison, K., 'History, representative institutions, and political rights in the French pre-Revolution, 1787–1789', *FHS* xv (1987), 68–98
Mathiez, A., *La Théophilanthropie et le culte décadaire 1796–1801: essai sur l'histoire religieuse de la Révolution*, Paris 1903
—— *La Vie chère et le mouvement social sous la terreur*, [Paris] 1927
Mayaud, J.-L., *Besaneon horloger, 1793–1914*, Besaneon 1994
Mayr, O., *Authority, liberty and automatic machinery in early modern Europe*, Baltimore 1986
Meinzer, M., *Der französische Revolutionskalender (1792–1805): Planung, Durchführung und Scheitern einer politischen Zeitrechnung*, Munich 1992
Menéres, C., 'L'Espace du temps: l'horloge', *Temps libre* ii (1981), 9–17
Meysonnier, S., *La Balance et l'horloge: la genèse de la penée libérale en France au XVIIIe siècle*, Paris 1989
Michelet, J. and G. Wright, *History of the French Revolution*, Chicago 1967
Moghadam, V. M., 'Gender and revolutions', in J. Foran (ed.), *Theorizing revolutions*, London 1997, 137–67
Monod-Casidy, H., 'Un Astronome-philosophe, Jérôme de Lalande', in T. Besterman (ed.), *Studies on Voltaire and the eighteenth century*, lvi, Geneva 1967, 907–30
Moran, G. T., 'Conceptions of time in early modern France: an approach to the history of collective mentalities', *Sixteenth Century Journal* xii (1981), 3–19
Morrell, J. B., 'Professors Robison and Playfair, and the *Theophobia Gallica*: natural philosophy, religion and politics in Edinburgh, 1789–1815', *Notes and Records of the Royal Society of London* xxvi (1971), 43–63
Muchembled, R., *Popular culture and elite culture in France, 1400–1750*, Baton Rouge 1985
Munn, N. D., 'The cultural anthropology of time: a critical essay', *Annual Review of Anthropology* xxi (1992), 93–123
Nash, G. B., 'The American clergy and the French Revolution', *William and Mary Quarterly* 3rd ser. xxii (1965), 393–7
Nesbet, R., *History of the idea of progress*, New York 1980
Nicolas, J., *La Révolution Franeaise dans les Alpes-Dauphiné et Savoie, 1789–1799*, Toulouse 1989
Nora, P., *Realms of memory: rethinking the French past*, New York 1996
Norbrook, D., *Writing the English Republic: poetry, rhetoric and politics, 1627–1660*, Cambridge 1999
O'Brien, K., *Narratives of enlightenment: cosmopolitan history from Voltaire to Gibbon*, Cambridge 1997
—— 'History and literature, 1660–1780', in J. Richetti (ed.), *Cambridge history of English literature, 1660–1780*, Cambridge 2005, 363–90
O'Connell, L. M, review of *Emulation*, H-France, Oct. 2006, http://www.h-net.org/reviews/showpdf.php?id=646

Ozouf, M., 'Space and time in the festivals of the French Revolution', *Comparative Studies in Society and History* xvii (1975), 372–84

—— *Festivals and the French Revolution*, Cambridge, MA 1988

—— 'Regeneration', in Furet and Ozouf, *Critical dictionary*, 781–91

—— 'Revolutionary Calendar', in Furet and Ozouf, *A critical dictionary*, 538–47

Painter, G. S., 'The idea of progress', *American Journal of Sociology* xxviii (1922), 257–82

Palmer, R. R., *The improvement of humanity: education and the French Revolution*, Princeton 1985

Pardaihlé-Galabrun, A., *La Naissance de l'intime: 3000 foyers parisiens, XVIIe–XVIIIe siècles*, Paris 1988

Peris, D., *Storming the heavens: the league of the Soviet godless*, Ithaca 1998

Perrie, M., D. C. B. Lieven, and R. G. Suny (eds), *Cambridge history of Russia*, Cambridge 2006

Perrot, M., *Les Ouvriers en grève*, Paris 1974

Petiteau, N., *L'Horlogerie des bourgeois bonquérants: histoire des établissements bourgeois de Damprichard Doubs, 1780–1939*, Besaneon 1994

Poole, R., '"Give us our eleven days!": calendar reform in eighteenth-century England', *Past & Present* cxlix (1995), 95–139

—— *Time's alteration: calendar reform in early modern England*, London 1998

Pouy, F., *Recherches sur les almanachs & calendriers*, Amiens 1874

Reichardt, R. and H.-J. Lüsebrink, 'Révolution à la fin du 18e siècle: pour une relecture d'un concept-clé du siècle des lumières', *Mots* xvi (1988), 35–68

Reinhard, M., *Le Département de la Sarthe sous le régime directorial*, Saint Brieuc 1936

Rey, A., *'Révolution': histoire d'un mot*, Paris 1989

Rey-Pailhade, J. de, 'Étude historique sur l'emploi du calendrier républicain et sur le temps décimal à Toulouse pendant la Révolution', *Bulletin de la Société de Géographie de Toulouse* iv (1908), 429–57

Riley, P. F., *A lust for virtue: Louis XIV's attack on sin in seventeenth-century France*, Westport 2001

Roche, D., 'Le Cabaret parisien et les manières de vivre du peuple', in M. Garden and Y. Lequin (eds), *Habiter la ville Xve–XXe siècles: actes de la tables ronde organisée avec l'aide de la D.G.R.S.T. et de la mission de la recherche urbaine*, Lyon 1984, 233–51

Rosenberg, D. and A. Grafton, *Cartographies of time: a history of the timeline*, New York 2010

Roussel, A., *Un Évêque assermenté, 1890–1802: Le Coz, évêque d'Ille-et- Vilaine, métropolitain du Nord-Ouest*, Paris 1899

Sagan, E., *Citizens and cannibals: the French Revolution, the struggle for modernity, and the origins of ideological terror*, Lanham, MD 2001

Saugera, E., *Bordeaux, port négrier: chronologie, économie, idéologie, XVIIe– XIXe siècles*, Biarritz 1995

Sauzay, J., *Histoire de la persécution révolutionnaire dans le departement du Doubs, de 1789 à 1801, d'apres les documents originaux inédits*, Besaneon 1867

Scattergood, J., 'Writing the clock: the reconstruction of time in the late Middle Ages', *European Review* xi (2003), 453–74
Schama, S., *Citizens: a chronicle of the French Revolution*, New York 1989
Schneider, R. A., *The ceremonial city: Toulouse observed, 1738–1780*, Princeton, NJ 1995
Sciout, L., *Histoire de la constitution civile du clergé, 1790–1801*, Paris 1872
Seifert, S., *Die Zeit schlägt ein neues Buch in der Geschichte auf: zum Französischen Revolutionskalender und zu seiner aufnahme in Deutschland*, Weimar 1989
Shaw, M., 'Reactions to the French republican calendar', *FH* xv (2001), 4–25
Shennan, J.H., *The parlement of Paris*, Ithaca 1968
Sherman, S., *Telling time: clocks, diaries, and English diurnal form, 1660–1785*, Chicago 1996
Shusterman, N., *Religion and the politics of time: holidays in France from Louis XIV through Napoleon*, Washington, DC 2010
Simon, R., 'La Révolution vécue par un bourg d'Ile-de-France: Beaune-en-Gatinais, 1789–1799 d'après les registres municipaux', *Revue d'histoire moderne et contemporaine* xviii (1971), 250–60
Skinner, G. W., 'Marketing and social structure in rural China', *Journal of Asian Studies* xxiv (1964), 11–16
Slavin, M., *The French Revolution in miniature: section Droits-de-l'Homme, 1789–1795*, Princeton, NJ 1984
Smith, M. M., *Mastered by the clock: time, slavery, and freedom in the American South*, Chapel Hill 1997
Soboul, A., *The French Revolution, 1787–1799: from the storming of the Bastille to Napoleon*, New York 1974
Sonenscher, M., 'Work and wages in Paris in the eighteenth century', in M. Berg, P. Hudson and M. Sonenscher (eds), *Manufacture in town and country before the factory*, Cambridge 1983, 147–72
Sonnet, M., 'Les Almanachs politiques parus pendant la Révolution Franeaise', *Bulletin des Bibliothèques de France* xxv (1980), 5–10
Spang, R. L., 'Paradigms and paranoia: how modern is the French Revolution?', *AHR* cviii (2003), 119–47
Steinberg, S. H., *Five hundred years of printing*, Harmondsworth 1955
Les Stroganoff: une dynastie de mécènes, Paris 2002
Suagnieux, J., 'Le Temps, l'espace et la presse au siècle des lumières', *Cahiers d'histoire* xxiii (1978), 313–34
Tackett, T., *Becoming a Revolutionary: the deputies of the French national assembly and the emergence of a Revolutionary culture, 1789–1790*, Princeton, NJ 1996
—— *When the king took flight*, Cambridge, MA 2003
Taylor, G., *The French Revolution and the London stage, 1789–1805*, Cambridge 2000
Tchoudinov, A., 'Les Papiers de Gilbert Romme aux archives russes', *AHRF* cccvi (1996), 257–65
Thomas, J., *Les Temps des foires: foires et marchés dans le Midi toulousain de la fin de l'ancien régime à 1914*, Toulouse 1993

Thompson, E. P., 'Time, work-discipline, and industrial capitalism', *Past & Present* xxxviii (1967), 56–97

—— *Customs in common*, London 1991

Tierder, I., 'La Calendrier républicain et ses incidences littéraires', *AHRF* ccxcii (1993), 259–67

Tilly, C., *The Vendee*, Cambridge, MA 1964

Ultée, J. M., 'The suppression of fêtes in France, 1666', *Catholic Historical Review* lvii (1976), 181–99

Van Kley, D., 'From the lessons of French history to truths for all times and all people: the historical origins of an anti-historical declaration', in D. Van Kley (ed.), *The French idea of freedom: the Old Regime and the Declaration of Rights of 1789*, Stanford 1994, 72–113

Voth, H.-J., 'Time and work in eighteenth-century London', *Journal of Economic History* lviii (1998), 29–58

—— *Time and work in England, 1750–1830*, Oxford 2000

Vovelle, M., *Les Métamorphoses de la fête en Provence de 1750 à 1820*, Paris 1976

—— *Religion et révolution: la déchristianization de l'an II*, Paris 1976

—— *La Révolution Franeaise: images et récit, 1789–1799*, iv, Paris 1986

—— *The Revolution against the Church: from reason to the supreme being*, Columbus, OH 1996

Vries, J. de, 'Peasant demand patterns and economic development: Friesland, 1550–1750', in W. N. Parker and E. L. Jones (eds), *European peasants and their markets: essays in agrarian economic history*, Princeton, NJ 1975, 205–68

—— 'The Industrial Revolution and the Industrious Revolution', *Journal of Economic History* liv (1994), 249–70

Walsh, L., '"Arms to be kissed a thousand times": reservations about lust in Didero's art criticism', in G. Perry and M. Rossington (eds), *Femininity and masculinity in eighteenth-century art and culture*, Manchester 1994, 162–83

Wauters, E., *Une Presse de province pendant la Révolution Franeaise: journaux et journalistes normands, 1785–1800*, Paris 1993

Weber, C., 'Freedom's missed moment', *Yale French Studies* xi (2001), 9–31

Weber, E., 'About Thermidor: the oblique uses of a scandal', *FHS* xvii (1991), 330–42

Whatmore, R., *Republicanism and the French Revolution: an intellectual history of Jean-Baptiste Say's political economy*, Oxford–New York 2000

Whipp, R., '"A time to every purpose": an essay in time and work', in P. Joyce (ed.), *The historical meanings of work*, Cambridge 1987, 210–36

Whitney, E., *Paradise restored: the mechanical arts from antiquity through the thirteenth century*, Philadelphia 1990

Williams, A., *The police of Paris, 1718–1789*, Baton Rouge 1979

Williams, R. C., 'The Russian Revolution and the end of time: 1900–1940', *Jahrbücher für Geschichte Ost europas* n.s. xliii (1995), 364–401

Woloch, I., *The new regime: transformations of the French civic order, 1789– 1820s*, New York 1994

Zerubavel, E., 'The French republican calendar: a case study in the sociology of time', *American Sociological Review* xlii (1977), 868–77
—— *Hidden rhythms: schedules and calendars in social life*, Chicago 1981
Zupko, R. E., *Revolution in measurement: western European weights and measures since the age of science*, Philadelphia 1990

Unpublished dissertations and papers

Arnould, F., 'Laurent Mégevand et l'introduction de l'horlogerie à Besaneon, 1793–1814', MA, Besaneon 1989
Friguglietti, J., 'The social and religious consequences of the French Revolutionary Calendar', Phd, Cambridge, MA 1966
Nelson, W., 'The weapon of time: constructing the future in France, 1750 to Year I', PhD, Los Angeles 2006
Perovic, S., 'Untameable time', PhD, Berkeley 2004
—— 'Between the volcano and the sun: the revolutionary calendar and Sylvain Maréchal', unpubl. paper, Modern France Workshop, University of Chicago, 25 Feb. 2005
Shaw, M., 'Time and the French Revolution, 1789–Year XIV', PhD, York 2000
Sherman, S. J., 'Telling time: clocks and calendars, secrecy and self-recording in English diurnal form, 1660–1795', PhD Columbia 1990
Shusterman, N., 'Festivals, calendars, and the nationalization of time in France, 1642–1815', PhD, Berkeley 2004
Tardieu, S., 'La Vie domestique dans le Mâconnais rural préindustriel', PhD, Paris 1964
Taws, R., 'Circulation and authority: print culture and the fabrication of political identity in revolutionary Paris', PhD, London 2005

Index

Printed and bound by CPI Group (UK) Ltd, Croydon, CR0 4YY
16/09/2026
14963452-0001